面向东盟职业教育双语新形态教材

网络互联技术（双语版）

Technology of Network Interconnection（Bilingual edition）

主　编　钟文基　邓启润　劳　薇
副主编　邓丽萍　周　顺　区倩如
　　　　张志秀　陈前军
参　编　石　巍　董其才　莫年发
　　　　吴丽萍　黄彩莹　陆　腾
　　　　颜　靖

北京理工大学出版社
BEIJING INSTITUTE OF TECHNOLOGY PRESS

内 容 简 介

本书以思科公司 Cisco Packet Tracer 模拟器为主要平台，以最新版 CCNA7.0 大纲为参考，以实验为基础，以项目式的任务为驱动，按网络工程行业的实际技术需求组织完整的知识框架，旨在打造简单易学且实用性强的轻量级计算机网络互联技术教程。

本书分 5 个项目共 20 个任务。项目 1 为认识计算机网络，描述如何部署一个畅通的网络，让计算机之间能相互通信；项目 2 为拓展网络，介绍怎样实现逻辑网络内部通信及 WLAN 部署；项目 3 为接入互联网，介绍了怎样实现企业内部网络和外部网络之间的通信；项目 4 为解析网络通信，介绍 OSI 参考模型各层协议、功能及工作原理；项目 5 为排除网络故障，讲解故障排除任务。完成本书的学习后，学生可以适应行业内绝大部分不同厂商的网络应用环境。

除了技术，本书还融入了课程思政的元素。将技术拓展到学习和生活中，帮助学生树立正确的人生观和世界观，培养思考的意识。

这本书在技术上没有任何门槛的要求。可以作为职业院校、应用型本科计算机网络技术相关课程教材用书。同时，对于从事网络管理和维护的技术人员，也可用于日常的技术参考，帮助他们系统地思考计算机网络运行原理。

版权专有　侵权必究

图书在版编目（CIP）数据

网络互联技术：英汉对照 / 钟文基，邓启润，劳薇主编. -- 北京：北京理工大学出版社，2024.4
ISBN 978-7-5763-3903-1

Ⅰ.①网… Ⅱ.①钟…②邓…③劳… Ⅲ.①互联网络-英、汉 Ⅳ.①TP393.4

中国国家版本馆 CIP 数据核字（2024）第 089498 号

责任编辑：陈莉华		**文案编辑**：李海燕		
责任校对：周瑞红		**责任印制**：施胜娟		

出版发行 /	北京理工大学出版社有限责任公司
社　　址 /	北京市丰台区四合庄路 6 号
邮　　编 /	100070
电　　话 /	（010）68914026（教材售后服务热线）
	（010）68944437（课件资源服务热线）
网　　址 /	http://www.bitpress.com.cn
版 印 次 /	2024 年 4 月第 1 版第 1 次印刷
印　　刷 /	河北盛世彩捷印刷有限公司
开　　本 /	787 mm×1092 mm　1/16
印　　张 /	17.5
字　　数 /	400 千字
定　　价 /	65.00 元

图书出现印装质量问题，请拨打售后服务热线，负责调换

前言

现在几乎没有人去讨论计算机网络的重要性,因为根本无法想象没有网络的世界会是什么样的。

本书以任务驱动的方式,先提出需求再解决问题,由浅入深引导读者一边实践一边思考。整本书的内容在多年教学实践的基础上,按人们的认知普遍规律,经过精心编排,做到"做中学、学中做"。

1. 循序渐进方式的内容编排

项目 1:认识计算机网络。主要是讲解计算机网络的要素及其发展史。通过使用 Cisco Packet Tracer 搭建一个最简单的只有 2 个节点的网络系统,配好 IP 地址,让学员能在最开始的时候,就能体会到什么是网络通信。这种快速而直观的收获能为后续的学习提供更大的动力。

项目 2:拓展网络。首先是实现了一个逻辑网络内部的数据通信,再拓展到不同逻辑网络之间的通信。加深理解 IP 地址、子网掩码的概念,同时理解网关的作用。其次是从有线接入拓展到无线接入的方式,进行了 WLAN 的部署。

项目 3:接入互联网。前面的内容只是关注一个企业内部网络的通信,而本项目关注的是企业内部网络和外部网络之间的通信实现。其解决的是内外网之间的通信问题,这一问题的解决,基本上让学员理解一般情况下人们访问互联网资源的过程。

项目 4:解析网络通信。这部分是本书的核心知识点之一,主要是基于 OSI 参考模型解释计算机网络的数据通信过程是怎样的。与其他教材不一样的是,这本书是运用 OSI 参考模型去解释数据通信过程,而不只是讲解 OSI 参考模型各层的功能。同时通过 OSI 参考模型理解交换机、路由器的工作机制,以及传输层 TCP 和 UDP 的工作原理。

项目 5:排除网络故障。主要讲解排除计算机网络故障的一般思路及经验总结。让学生通过 2 个故障排除任务及思考与训练部分的任务练习,灵活运用 OSI 参考模型的一般思路来排除网络故障,同时积累经验。

2. 尝试融入课程思政元素

在编写的过程中，教材的定位经历了两次调整：一开始只是想写一本教材，能把计算机网络的灵魂写出来就可以，配套微视频、PPT 等资源，方便开展教学。再后来有专家建议我们如果能将课程思政的因素考虑进去，更符合当前形势下教学的需求，于是我们也把课程思政作为编写本书的一个目标。

本书并没有将思政内容硬生生地切入计算机网络技术，而是尝试在讲解技术的过程中进行扩展，有机融入思政元素，感觉更自然。

3. 因材施教运用于不同学习对象

本书适用于应用型本科、高职以及培训机构计算机网络基础类课程，可根据课时的长短以及专业需要对内容进行取舍。

完成本书的学习，建议安排不少于 75 学时。但对于不同的教学对象可以做出内容和学时的调整。由于客观条件的限制，如果书本的内容没有办法在课堂上完成，则其他没有讲授的内容可以给学习能力好的学生提供更大的学习空间，做到因材施教。

由于时间仓促，加上我们水平有限，书中难免有不妥和错误之处，恳请同行专家指正。并将意见和建议发送 E-mail：redhat70@163.com，谢谢。

编　者

Preface

Today, no one questions the importance of computer networks, for it is impossible to imagine a world without them.

This book adopts a task-driven approach, raising needs before solving problems, and gradually guides readers from the simple to the complex, encouraging them to think while practicing. The content of the entire book, based on years of teaching experience and crafted in line with general cognitive rules, follows the principle of "learning by doing."

1. Content Arrangement in a Progressive Manner

Project 1: Understanding Computer Networks. This project primarily explains the elements and the history of computer networks. Using Cisco Packet Tracer, students build a simple two-node network system and configure IP addresses, allowing them to experience network communication right from the start. This quick and intuitive gain provides greater motivation for subsequent learning.

Project 2: Expanding the Network. It starts with data communication within a logical network and then expands to communication between different logical networks. This deepens the understanding of IP addresses, subnet masks, and the role of gateways. It also includes the deployment of WLAN, transitioning from wired accesses to wireless ones.

Project 3: Accessing the Internet. This focuses on the communication between internal and external networks of an enterprise, addressing the issue of internal and external network communication.

Project 4: Analyzing Network Communication. Based on the OSI reference model, this section is one of the core knowledge points of the book, explaining the data communication process in computer networks by using the OSI reference model.

Project 5: Troubleshooting Network Faults. This mainly discusses general strategies and experiences in troubleshooting network faults, allowing students to apply the OSI reference model to diagnose and fix network issues.

2. Incorporating Course Ideological and Political Elements

The book's orientation has undergone two adjustments during the writing process. Initially, the goal was to create a textbook that encapsulates the essence of computer networking. Later, experts suggested incorporating course ideological and political elements, aligning better with current educational needs. This book naturally integrates these elements into the technical explanations.

3. Tailoring Teaching Methods to Different Learners

The book is suitable for applied undergraduates, vocational colleges, and computer network basic courses at training institutions. Depending on the course length and specialty, content can be adapted accordingly. The book is designed for at least 75 hours of study but can be adjusted for different teaching objectives. Uncovered content in class can provide more learning opportunities for capable students.

Due to time constraints and limitations, there might be inaccuracies in the book. We welcome corrections and suggestions from our fellow teachers and experts, and our E-mail is: redhat70@163.com. Thank you.

The Editors

目 录

项目 1　认识计算机网络 ·· 1

　　任务 1：初识计算机网络——Internet 简史 ·································· 3
　　任务 2：搭建最简单的网络拓扑——2 个节点的网络 ······················ 5
　　任务 3：为计算机配置最简单的参数——IP 地址 ···························· 8
　　小结与拓展 ·· 10
　　思考与训练 ·· 16

Project 1　Understanding Computer Networks ···································· 18

　　Task 1：Introduction to Computer Networks—A Brief History of the Internet ········· 20
　　Task 2：Building the Simplest Network Topology—A Two-Node Network ············· 23
　　Task 3：Configuring the Simplest Parameters for Computers—IP Addresses ········· 26
　　Summary and Expansion ·· 28
　　Thinking and Training ··· 35

项目 2　拓展网络 ·· 36

　　任务 1：构建小型局域网络——交换机 ··· 37
　　任务 2：确定计算机所在的网络——子网掩码 ································ 39
　　任务 3：用路由器连接不同的网络——网关 ··································· 43
　　任务 4：用三层交换机连接不同的网络——SVI（VLAN、Trunk） ······· 51
　　任务 5：部署无线局域网——WLAN ·· 60
　　小结与拓展 ·· 64
　　思考与训练 ·· 72

Project 2　Expanding the Network ·· 73

　　Task 1：Building a Small Local Area Network—Switch ··························· 74

1

Task 2: Identifying the Network of a Computer—Subnet Mask 76
Task 3: Connecting Different Networks with a Router—Gateway 80
Task 4: Connecting Different Networks by Using Layer 3 Switches—SVI
（VLAN, Trunk） 88
Task 5: Deploy Wireless Local Area Network—WLAN 97
Summary and Expansion 102
Thinking and Training 110

项目 3　接入互联网 112

任务 1：将数据发送到边界路由器——静态路由 115
任务 2：实现内网用户访问互联网服务器——端口地址转换（PAT） 120
任务 3：实现互联网用户访问内网主机——网络地址转换（静态 NAT） 126
任务 4：减少路由表的路由条目——路由汇总（默认路由） 129
任务 5：部署域名服务器——域名服务（DNS） 134
小结与拓展 137
思考与训练 142

Project 3　Access the Internet 143

Task 1: Send Data to Border Routers—Static Routes 147
Task 2: Enabling Intranet Users to Access Internet Servers—Port
Address Translation (PAT) 152
Task 3: Enabling Internet Users to Access Intranet Hosts—Network Address Translation
(Static NAT) 158
Task 4: Reducing Route Entries in the Routing Table—Route Summarisation
(Default Route) 161
Task 5: Deploying Domain Name Servers—Domain Name Service (DNS) 168
Summary and Expansion 171
Thinking and Training 177

项目 4　解析网络通信 178

任务 1：初识 OSI 参考模型 179
任务 2：解析一次简单的数据通信过程——OSI 各层的协作 182
任务 3：交换机参与数据通信——数据链路层 185
任务 4：路由器参与数据通信——网络层 187
任务 5：TCP 确保可靠的数据传输——传输层 189
小结与拓展 193
思考与训练 202

Project 4　Analyzing Network Communications ……… 203

Task 1：A First Look at the OSI Reference Model ……… 204
Task 2：Parsing a Simple Data Communication Process—Collaboration of
　　　　OSI Layers ……… 209
Task 3：Switch Participation in Data Communications—Data Link Layer ……… 212
Task 4：Router Participation in Data Communications—Network Layer ……… 215
Task 5：TCP Ensures Reliable Data Transfer—Transport Layer ……… 217
Summary and Expansion ……… 222
Thinking and Training ……… 235

项目 5　排除网络故障 ……… 236

任务 1：排除网络第 2 层故障——Vlan、Trunk、SVI（交换机虚拟端口）……… 238
任务 2：排除网络第 3 层故障——路由追踪 Tracert、NAT ……… 243
小结与拓展 ……… 249
思考与训练 ……… 250

Project 5　Troubleshooting the network ……… 251

Task 1：Troubleshooting Network Layer 2—Vlan, Trunk, SVI
　　　　（Switch Virtual Ports）……… 253
Task 2：Troubleshooting Network Layer 3—Route Tracing tracert, NAT ……… 259
Summary and Expansion ……… 266
Thinking and Training ……… 267

参考文献 ……… 269

项目 1

认识计算机网络

作为学习的起点，我们将在本项目弄清楚什么是计算机网络，它是怎样诞生并发展起来的。我们开始尝试用模拟器创建一个最简单的能通信的计算机网络模型。完成本项目的学习后，可以对计算机网络及其通信过程有一个相对直观的感受，同时提升后续学习的信心。

知识目标

（1）理解什么是计算机网络；
（2）了解计算机网络发展历史；
（3）理解计算机网络常见的拓扑结构；
（4）理解 IP 地址的概念及其作用。

技能目标

（1）学会用 Cisco Packet Tracer 构建简单的计算机网络模型；
（2）掌握为计算机配置 IP 地址的方法。

相关知识

1. 计算机网络是什么

"网络"的英文叫"network"，表示网状系统、关系网、人际网、相互关系（或配合）的系统、（互联）网络。

"网络"一词在不同的领域表达的意思可以是不一样的。在我们生活中用来表达比较多的有：有线电视网络、电话网络、移动通信网络，甚至可以指人际关系网络。但是在这里，我们主要探讨的是计算机之间的通信，也就是计算机网络。

那什么是计算机网络呢？行业内普遍认为，一组相互独立但彼此连接的计算机集合就可以称为计算机网络系统。从这个描述，我们可以了解到计算机网络就是为了实现两个目标：一个是独立又相互连接的计算机之间相互传递信息；另一个是实现计算机之间的资源共享。

2. 计算机网络的分类

计算机网络按覆盖的范围大小可以分为以下三类：

局域网（Local Area Network，LAN）：针对较小地理区域内的用户和终端设备提供访问的网络基础设施，通常是由个人或 IT 部门拥有并管理的企业、家庭或小型企业网络。

城域网（Metropolitan Area Network，MAN）：覆盖的物理区域大于 LAN 但小于 WAN 的网络基础设施（例如一个城市）。MAN 通常由单个实体（如大型组织）运营。

广域网：（Wide Area Network，WAN）：针对广泛地理区域内的其他网络提供访问的网络基础设施，通常由通信服务提供商拥有并管理。

3. 局域网的拓扑结构

局域网的拓扑分为物理拓扑和逻辑拓扑两种，它们所表示的信息是不一样的。物理拓扑描述的是如何将设备用线缆物理地连接在一起，而逻辑拓扑描述的是设备之间如何通过物理拓扑进行通信。

物理拓扑定义了计算机终端系统的物理互连方式。在共享介质 LAN 上，终端设备可以使用以下物理拓扑互连：

（1）星形：将终端设备连接到中心的网络设备。早期的星形拓扑使用以太网集线器互连终端设备。由于集线器的通信效率极低，因此现在星形拓扑使用以太网交换机。星形拓扑安装简易、扩展性好（易于添加和删除终端设备），而且容易排除网络故障，如图 1.1 所示。

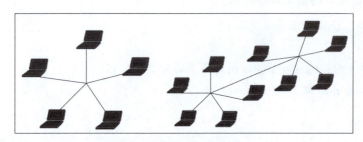

图 1.1 星形和拓展星形拓扑

（2）扩展星形：一台以太网交换机的端口是有限的，满足不了更多用户接入，需要将交换机互连，以拓展出更多端口，供更多用户的接入需求。在扩展星形拓扑中，额外的以太网交换机与其他星形拓扑互连。扩展星形是一种混合拓扑。

（3）总线：所有终端系统都相互连接，并在总线两端各自挂载一个 50 Ω 的电阻吸收电磁信号。终端设备互连时不需要基础设施设备（例如集线器、交换机）。因为总线拓扑价格低廉而且安装简易，所以传统的以太网络中会采用同轴电缆的总线拓扑。

（4）环形：终端系统与其各自的邻居相连，形成一个环状。与总线拓扑不同，环形拓扑不需要端接。环形拓扑用于传统的光纤分布式数据端口（Fiber Distributed Data Interface，FDDI）和令牌环（Token Ring）网络。如图 1.2 所示是总线拓扑和环形拓扑。

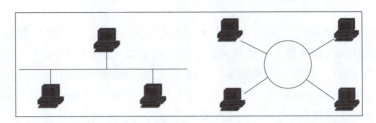

图 1.2 总线拓扑和环形拓扑

任务 1：初识计算机网络——Internet 简史

任务目标：了解计算机网络的诞生及发展历史。

1. 国际互联网发展历史

初识计算机网络

当今最大的广域网叫因特网（Internet），覆盖全球。我们平时所表达的"上网"，指的就是访问因特网。其发展历史可以追溯到 20 世纪 50 年代末期。

在 20 世纪 50 年代，通信研究者认识到需要允许在不同计算机用户和通信网络之间进行常规的通信。这促使了分散网络、排队论和分组交换的研究。

1960 年，美国国防部高等研究计划署（Advanced Research Projects Agency，ARPA）出于冷战考虑创建的 ARPANET 引发了技术进步，并使其成为互联网发展的中心。ARPANET 的发展始于 2 个网络节点［由伦纳德·克莱因罗克带领的加利福尼亚大学洛杉矶分校的网络测量中心与加利福尼亚州门罗帕克斯坦福国际研究院（Stanford Research Institute，SRI）道格拉斯·恩格尔巴特的 NLS 系统］之间的连接。加入 ARPANET 的第 3 个节点是加利福尼亚大学圣塔芭芭拉分校，第 4 个节点是犹他大学。到 1971 年底，已经有 15 个节点连接到 ARPANET。

1973 年 6 月，挪威地震数组所（Norwegian Seismic Array，NORSAR）连接到 ARPANET，成为美国本土之外的第一个网络节点。

1974 年，罗伯特·卡恩和文顿·瑟夫提出 TCP/IP 协议，定义了在计算机网络之间传送报文的方法（他们在 2004 年也因此获得图灵奖）。

1986 年，美国国家科学基金会创建了超级计算机中心与学术机构之间基于 TCP/IP 技术的骨干网络 NSFNET，速度最初为 56 Kbit/s，接着为 T1（1.5 Mbit/s），最后发展至 T3（45 Mbit/s）。

商业互联网服务提供商（Internet Service Provider，ISP）出现于 1980 年代末和 1990 年代初。

1990 年，ARPANET 退役。

20 世纪 80 年代中后期，互联网在欧洲和澳大利亚迅速扩张，并于 20 世纪 80 年代后期和 1990 年代初期扩展至亚洲。

1989 年中，MCI Mail 和 CompuServe 与互联网创建了连接，并且向 50 万大众提供了电子邮件服务。

1990年3月，康奈尔大学和欧洲核子研究中心之间架设了NSFNET和欧洲之间的第一条高速T1（1.5 Mbit/s）连接。6个月后，蒂姆·伯纳斯-李编写了第一个网页浏览器。

到1990年圣诞节，蒂姆·伯纳斯-李创建了运行万维网所需的所有工具：超文本传输协议（Hypertext Transfer Protocol，HTTP）、超文本标记语言（Hypertext Markup Language，HTML）、第一个网页浏览器、第一个网页服务器和第一个网站。

1995年，NSFNET退役时，互联网解除了最后的商业流量限制，在美国完全商业化。

2. 中国互联网发展历史

中国互联网是全球第一大网，用户数量最多，联网区域最广。但是我国互联网整体发展时间比较短。

1986年8月25日，瑞士日内瓦时间4点11分，北京时间11点11分，由当时任高能物理所ALEPH组组长的吴为民，从北京发给ALEPH的领导——位于瑞士日内瓦西欧核子中心的诺贝尔奖获得者斯坦伯格（Jack Steinberger）的电子邮件（E-mail）是中国第一封国际电子邮件。

1989年8月，中国科学院负责建设"中关村教育与科研示范网络"（NCFC）。

1994年4月，NCFC与美国NSFNET直接互联，实现了中国与Internet全功能网络连接，标志着我国最早的国际互联网络的诞生。

1994年，中国第一个全国性TCP/IP互联网：CERNET示范网工程建成。

1998年，CERNET研究者在中国首次搭建IPv6试验床。

2000年，中国三大门户网站搜狐、新浪、网易在美国纳斯达克挂牌上市。

2001年，下一代互联网地区试验网在北京建成验收。

2003年，下一代互联网示范工程CNGI项目开始实施。

2020年3月，我国网民规模为9.04亿人，其中手机网民规模为8.97亿人，我国网民使用手机上网的比例达99.3%；农村网民规模为2.55亿人，占整体网民的28.2%；网络购物用户规模为7.10亿人，网络支付用户规模为7.68亿人。

3. 网络空间主权

互联网创造了人类生活新空间，自然也拓展了国家治理新领域。随着信息技术革命的日新月异，互联网对国际政治、经济、文化、社会、军事等领域的发展产生了深刻影响。信息化和经济全球化相互促进，互联网日益成为创新驱动发展的先导力量，融入社会生活的方方面面，深刻改变了人们的生产和生活方式，有力推动着社会发展。它是信息自由流动的主脉，意见自由交流的论坛，志趣相投者的社区，事业经营者的阵地，市场腾飞的翅膀，国家繁盛的支点。互联网真正让世界变成了地球村，让国际社会越来越成为你中有我、我中有你的命运共同体。继陆、海、空、天之后，网络空间已经成为人类生产生活的第五疆域。

2015年7月1日，十二届全国人大常委会第十五次会议通过了《中华人民共和国国家安全法》。新的国安法规定，国家建设网络与信息安全保障体系……加强网络管理，防范、制止和依法惩治网络攻击、网络入侵、网络窃密、散布违法有害信息等网络违法犯罪行为，维护国家网络空间主权、安全和发展利益。首次明确了"网络空间主权"的概念。

网络空间主权，是指一个国家在建设、运营、维护和使用网络，以及在网络安全的监督管理方面所拥有的自主决定权。网络空间主权是国家主权在网络空间中的自然延伸和表现，是国家主权的重要组成部分。作为国家主权的延伸和表现，网络空间主权集中体现了国家在网络空间可以独立自主地处理内外事务，享有在网络空间的管辖权、独立权、自卫权和平等权等权利。

4. 网络安全法

随着互联网的快速发展，各种网络安全问题也接踵而来：网络入侵、网络攻击等非法活动威胁信息安全；非法获取公民信息、侵犯知识产权、损害公民合法利益；宣扬恐怖主义、极端主义，严重危害国家安全和社会公共利益。

2016年11月7日通过的《中华人民共和国网络安全法》为网络安全、维护网络空间主权和国家安全、社会公共利益，保护公民、法人和其他组织的合法权益，促进经济社会信息化健康发展提供了法制保障。

我国的网络安全法明确了八个主要方面的内容：
①将信息安全等级保护制度上升为法律；
②明确了网络产品和服务提供者的安全义务和个人信息保护义务；
③明确了关键信息基础设施的范围和关键信息，基础设施保护制度的主要内容；
④明确了国家网络信息部门对网络安全工作的统筹协调职责和相关监督管理职责；
⑤明确网络实名制，并明确了网络运营者对公安机关、国家安全机关维护网络安全和侦查犯罪的活动提供技术支持和协助的义务；
⑥进一步完善了网络运营者收集使用个人信息的规则，积极保护个人网络安全的义务与责任；
⑦明确建立国家统一的监测预警信息通报和应急处置制度和体系；
⑧对支持促进网络安全发展的措施作了规定。

网络安全法的实施有效促进了我国互联网的健康发展。其有效遏制了出售个人信息、网络诈骗等行为；以法律形式明确"网络实名制"、保护关键信息基础设施、惩治攻击破坏我国关键信息基础设施的境外组织和个人；当发生重大突发事件时，可采取"网络通信管制"。

课堂练习：对于如何遵守我国的网络安全法，谈一下自己的想法。

任务2：搭建最简单的网络拓扑——2个节点的网络

任务目标：用Cisco Packet Tracer搭建一个2个节点的计算机网络。

下面我们用思科公司的模拟器Cisco Packet Tracer（version 7.3.x）搭建一个最简单的计算机网络拓扑，一个只有2个节点的网络。

步骤1：到思科网络技术学院的官网下载Cisco Packet Tracer最新的版本，并安装。

步骤2：运行Cisco Packet Tracer，按要求输入思科网络技术学院的账号和密码后，进

搭建简单的网络拓扑

入工作界面，如图 1.3 所示。

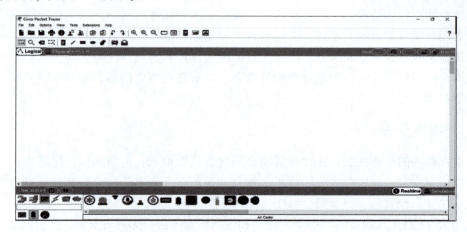

图 1.3　Cisco Packet Tracer 工作界面

思科网络技术学院的账号，可以通过上课的老师创建。如果没有账号，可以在登录界面单击右下角的"Guest Login"按钮用 guest 身份登录，如图 1.4 所示。

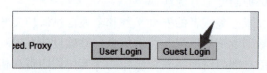

图 1.4　User Login 和 Guest Login 登录

步骤 3：在工作界面的左下角选择"End Device"（终端设备）图标，然后在出现的设备中，选择第一台"PC"（个人计算机），在工作区单击一下，放置一台 PC，如图 1.5 所示。

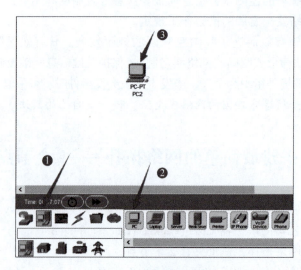

图 1.5　选择终端设备

步骤 4：用同样的方法，再添加第二台 PC，如图 1.6 所示。

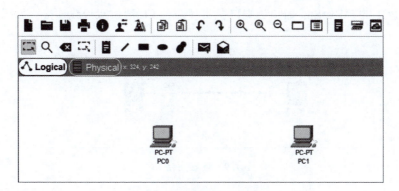

图 1.6 添加第二台 PC

步骤 5：选择介质 Copper Cross-Over（交叉双绞线），如图 1.7 所示。

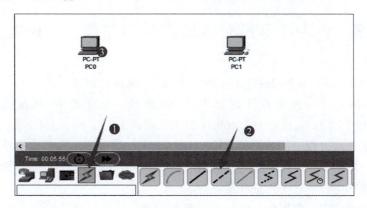

图 1.7 选择介质

步骤 6：用交叉双绞线把两台 PC 机连在一起。

用鼠标左键单击 PC0，在出现的菜单中选择"FastEthernet0"（快速以太网 0 号端口），如图 1.8 所示。

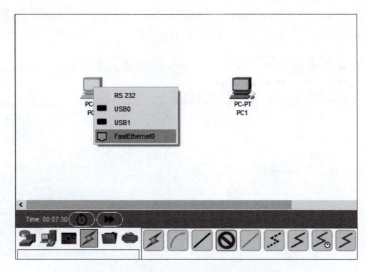

图 1.8 将双绞线连接到快速以太网端口

再选择 PC1 的"FastEthernet0"端口,实现两台 PC 通过双绞线连接,如图 1.9 所示。

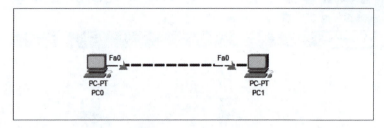

图 1.9　连接两台 PC

以上就是最简单的计算机网络的拓扑结构——两台计算机通过交叉双绞线直连。

课堂练习:用 Cisco Packet Tracer 模拟器搭建一个最简单的计算机网络拓扑。

任务 3:为计算机配置最简单的参数——IP 地址

任务目标:掌握 Cisco Packet Tracer 中计算机 IP 地址的配置方法(在本项目任务 1 的拓扑中,为两台计算机配置 IP 地址:PC0 的 IP 地址为 192.168.0.1,PC1 的 IP 地址为 192.168.0.2。

为计算机配置最简单的参数——IP 地址

计算机网络通信过程中的数据包传输,类似于我们发快递。我们去寄快递的时候,是一定要填一个发件人地址和收件人地址的,否则快递员就不知道要把快递送到哪里——这个地址非常重要。

计算机网络通信中的数据包一样,也需要类似这样的地址,我们把它叫做 IP 地址。类似的,为了能很好地把数据包从发送方传到接收方,IP 数据包上需要两个 IP 地址:发送方的 IP 地址(源 IP 地址)和接收方的 IP 地址(目的 IP 地址)。

也就是说,网络上的计算机,不管是发送方还是接收方,必须要有一个 IP 地址才可以参与网络通信。那如何为计算机配置 IP 地址呢?

步骤 1:单击 PC0,在出现的配置对话框中选择"Desktop"(桌面)选项卡,在选项卡中选择"IP Configuration"(IP 配置),如图 1.10 所示。

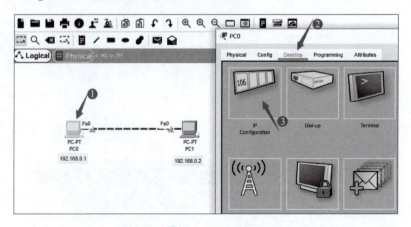

图 1.10　选择"IP Configuration"

步骤 2：在 IP Configuration 界面的"IP Address"（IP 地址）栏输入我们规划好的 IP 地址：192.168.0.1，如图 1.11 所示。

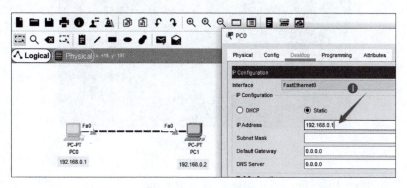

图 1.11　输入 IP 地址

注：配置完 IP 地址以后，下面的 Subnet Mask（子网掩码）会自动生成，这里先不去关注，下一个项目会进行关于子网掩码的学习。

关闭对话框退出。

步骤 3：用同样的方法，为 PC1 配置 IP 地址（192.168.0.2）。

步骤 4：通信测试。

如图 1.12 所示，依次单击"Desktop"和"Command Prompt"，启动 Command Prompt（命令提示符）工具，进入 CLI 界面。在 CLI 界面输入测试命令：ping 192.168.0.2，如图 1.13 所示。

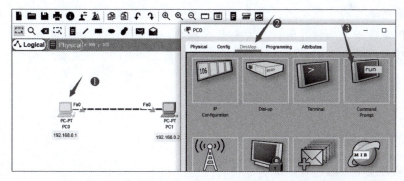

图 1.12　依次单击"Desktop"和"Command Prompt"

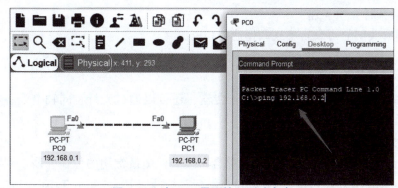

图 1.13　在 CLI 界面输入测试命令

按【回车】键，执行测试命令，结果如图 1.14 所示。

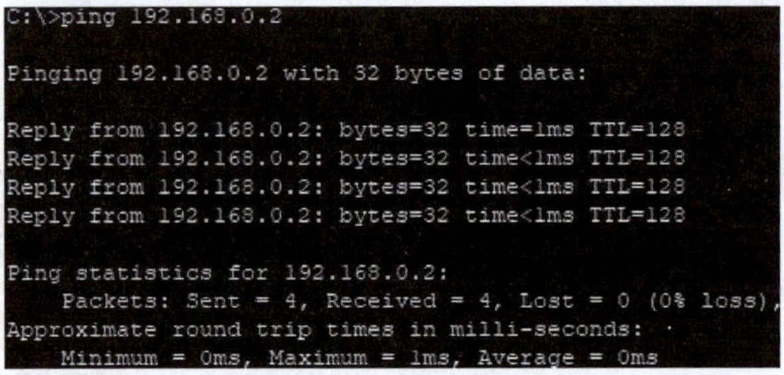

```
C:\>ping 192.168.0.2

Pinging 192.168.0.2 with 32 bytes of data:

Reply from 192.168.0.2: bytes=32 time=1ms TTL=128
Reply from 192.168.0.2: bytes=32 time<1ms TTL=128
Reply from 192.168.0.2: bytes=32 time<1ms TTL=128
Reply from 192.168.0.2: bytes=32 time<1ms TTL=128

Ping statistics for 192.168.0.2:
    Packets: Sent = 4, Received = 4, Lost = 0 (0% loss),
Approximate round trip times in milli-seconds:
    Minimum = 0ms, Maximum = 1ms, Average = 0ms
```

图 1.14　测试结果

看到"Reply from 192.168.0.2：bytes=32 time=1ms TTL=128"，说明 PC0 成功发送了一个数据包过去给 192.168.0.2（PC1），并且收到了 PC1 的应答（reply），说明 PC0 和 PC1 能正常通信。

IP 地址标记互联网中不同的主机，用于在数据通信中表达数据从哪里发出来，要发送到网络的哪个位置。类似人类社会的家庭住址一样，如果你要写信给一个人，你就要知道他（她）的地址，这样邮递员才能把信送到。

课堂练习：在任务 1 拓扑的基础上，配置两台计算机的 IP 地址，让两台计算机之间能相互 ping 通。

素质拓展：计算机获得合法的 IP 地址，就可以开始参与网络通信，也为人类打开了另外一扇通往世界的大门。但从本质上说，网络生活也是我们真实生活的扩展。大学生通过网络接触到前所未有的广阔空间，能更加有效和广泛地获取信息、学习知识、交流情感和了解社会。但是，一个人的时间和精力都是有限的，在网上消耗的时间多，在其他方面投入的时间就少。大学生应当合理安排上网时间，约束上网行为，避免沉迷网络。

小结与拓展

1. 构成计算机网络的要素

计算机网络技术应该包含以下 3 个要素：
（1）两台或者两台以上的计算机；
（2）通过某种介质互连；
（3）（遵守共同的协议）实现数据通信和资源共享。

2. 常用的传输介质

网络的数据除了可以通过双绞线进行传输，还可以通过光纤、同轴电缆、无线电波等介质传输。

（1）双绞线。

双绞线（Twisted Pair，TP）是一种综合布线工程中最常用的传输介质。双绞线分屏蔽双绞线（Shielded Twisted Pair，STP）和非屏蔽双绞线（Unshielded Twisted Pair，UTP）两类。计算机网络互联使用的是非屏蔽双绞线，是由 8 根具有绝缘保护层的 22～26 号铜导线组成的，分 4 组。每 2 根 1 组，按一定密度互相绞在一起，"双绞线"的名字也是由此而来。每根导线在传输中辐射出来的电波会被另 1 根线上发出的电波抵消，有效降低信号干扰的程度。

电气电子工程师协会（Institute of Electrical and Electronics Engineers，IEEE）定义了铜缆的电气特性。IEEE 按照它的性能对 UTP 布线划分等级。电缆分类的依据是它们承载更高速率带宽的能力。例如，5 类电缆通常用于 100BASE-TX 快速以太网安装。其他类别包括增强型 5 类电缆、6 类电缆和 6a 类电缆。

为了支持更高的数据传输速率，人们设计和构造了更高类别的电缆。随着新的千兆位以太网技术的开发和运用，如今已经很少采用 5e 类电缆，新建筑安装推荐使用 6 类电缆。

3 类电缆最初用于语音线路的语音通信，但后来用于数据传输。5 类和 5e 类电缆用于数据传输。5 类电缆支持 100 Mbit/s，5e 类电缆支持 1 000 Mbit/s，6 类电缆在每对线之间增加了一个分隔器以支持更高的速度，支持高达 10 Gbit/s。7 类电缆也支持 10 Gbit/s。8 类电缆支持 40 Gbit/s 的数据传输。

一些制造商制造的电缆超出了 TIA/EIA 6a 类电缆的规格，将其称为 7 类电缆。UTP 电缆的端头通常为 RJ45 连接器，如图 1.15 所示。

图 1.15　RJ45 连接器示意图

双绞线根据线序的不一样，分为交叉线和直通线两类。一般情况下，同种设备之间的连接用交叉线（双绞线一头的线序为 EIA/TIA-568A，另外一头为 EIA/TIA-568B），不同种设备之间的连接用直通线（双绞线两头的线序均为 EIA/TIA-568B）。

EIA/TIA-568A 和 EIA/TIA-568B 线序标准如表 1.1 所示。

表 1.1　EIA/TIA-568A 和 EIA/TIA-568B 线序标准

标准	序号对应的颜色							
	1	2	3	4	5	6	7	8
EIA/TIA-568A	绿白	绿	橙白	蓝	蓝白	橙	棕白	棕
EIA/TIA-568B	橙白	橙	绿白	蓝	蓝白	绿	棕白	棕

注：有一个情况比较特殊，就是计算机和路由器之间的连接用交叉线。

（2）光纤。

与其他网络介质相比，光纤能够以更远的距离和更高的带宽传输数据。不同于铜缆，光纤传输信号的衰减更少，并且完全不受电磁干扰或射频干扰的影响。光纤常用于互联网络设备。

光纤是一种由非常纯的玻璃制成的极细透明的弹性线束，和人的头发丝差不多细。通过光纤传输时，数据会被编码成光脉冲。光纤用作波导管或光导管，以最少的信号丢失来传输两端之间的光。

想象有一个空纸巾卷筒，其内部像是由一个镜子覆盖，长为 1 000 m，并且有一个小激光棒用于以光速发出莫尔斯电码信号，实质上这就是光缆运行的方式，只不过其直径更小，并且使用了复杂的光技术。

光纤通常分为两种类型：单模光纤（Single-Mode Fiber，SMF）和多模光纤（Multi-Mode Fiber，MMF）。

单模光纤包含一个极小的芯，使用昂贵的激光技术来发送单束光，如图1.16所示。单模光纤在跨越数百公里的长距离传输情况下很受欢迎，例如应用于长途电话和有线电视中的光纤。

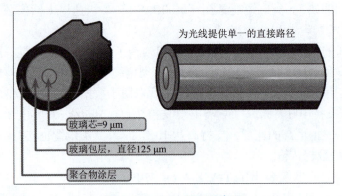

图1.16　SMF原理示意图

多模光纤包含一个稍大的芯，使用LED发射器发送光脉冲。具体而言，LED发出的光从不同角度进入多模光纤，如图1.17所示。普遍用于LAN中，因为它们可以由低成本的LED提供支持。它可以通过长达550 m的链路提供高达10 Gbit/s的带宽。

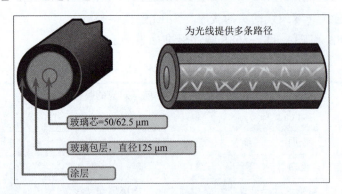

图1.17　MMF原理示意图

（3）无线电波。

无线介质使用无线电或微波频率来承载代表数据通信二进制数字的电磁信号。无线介质提供所有介质中最好的移动特性，而且启用无线的设备数量不断增加。无线现在是用户连接到家庭和企业网络的主要方式。

以下是无线网络的一些局限性：

①覆盖面积：无线数据通信技术非常适合开放环境。但是，在楼宇和建筑物中使用的某些建筑材料以及当地地形将会限制它的有效覆盖。

②干扰：无线电易受干扰，可能会受到家庭无绳电话、某些类型的荧光灯、微波炉和

其他无线通信装置等常见设备的干扰。

③安全性：无线通信覆盖无须进行介质的物理接线。因此，未获得网络访问授权的设备和用户可以访问传输。所以网络安全是无线网络管理的重要组成部分。

④共享介质：无线局域网（Wireless Local Area Network，WLAN）以半双工模式运行，意味着一台设备一次只能发送或接收。无线介质由所有无线用户共享。许多用户同时访问WLAN会导致每个用户的带宽减少。

无线数据通信的IEEE和电信行业标准包括数据链路层和物理层。常见的标准包括：

①WiFi（IEEE 802.11）：无线LAN（WLAN）技术，通常称为WiFi。WLAN使用一种称为"载波侦听多路访问/冲突避免（Carrier-Sense Multiple Access with Collision Avoidance，CSMA/CA）"的争用协议。无线NIC在传输数据之前必须先侦听，以确定无线信道是否空闲。如果其他无线设备正在传输，则NIC必须等待信道空闲。WiFi是WiFi联盟的标记。WiFi与基于IEEE 802.11标准的认证WLAN设备结合使用。

②蓝牙（IEEE 802.15）：这是一个无线个人局域网（Wireless Personal Area Network，WPAN）标准，通常称为蓝牙。它采用设备配对过程进行通信，距离为1~100 m。

③WiMAX（IEEE 802：16）：通常称为微波接入全球互通（World Interoperability for Microwave Access，WiMAX），这个无线标准采用点到多点拓扑结构，提供无线带宽接入。

④Zigbee（IEEE 802.15.4）：Zigbee是一种用于低数据速率、低功耗通信的规范。它适用于需要短距离、低数据速率和长电池寿命的应用。Zigbee通常用于工业和物联网（Internet of Things，IoT）环境，如无线照明开关和医疗设备数据采集。

3. 非屏蔽双绞线解决电磁干扰或射频干扰影响方法

UTP电缆并不使用屏蔽层来对抗电磁干扰或射频干扰的影响。相反，电缆设计者通过以下方式来减少串扰的负面影响：

（1）抵消：电缆设计者现在对电路中的电线进行配对。当电路中的两根电线紧密排列时，彼此的磁场正好相反。因此，这两个磁场相互抵消，也抵消了所有的外部EMI和RFI干扰信号。

（2）变化每个线对的扭绞次数：为了进一步增强配对电线的抵消效果，设计者会变化电缆中每个线对的扭绞次数。UTP电缆必须遵守精确的规定来管理每米电缆所允许的扭绞次数或编织数。请注意，橙色/橙白色线对比蓝色/蓝白色线对的扭绞次数要少。每个彩色线对扭绞的次数不同。

4. IP 地址

IP地址是主机在网络中的一个标识。

根据协议版本不一样，分为IPv4地址和IPv6地址，这里讨论的是IPv4地址。IPv4地址由32位二进制组成，一般我们将其用3个点号分4段，并将这4段数字分别转换成十进制来表达，提高可读性，如图1.18所示。

IP地址由网络位部分和主机位部分组成。网络位部分标记的是该IP地址所在的网络的编号，主机位部分是拥有该IP地址的主机在网络中的编号。我们可以把互联网类比成一个城市或者更大的区域，那么一个小网络可以类比成一条街，网络位部分就是这条街的编号，

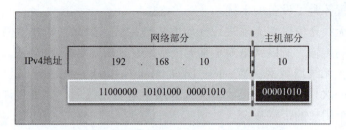

图 1.18　IPv4 地址的十进制和二进制表达方式

而主机位部分就是这条街上每个单位的门牌号。网络设备就是通过识别 IP 地址的网络位部分找到目标网络，然后在这个网络内找到目标主机。就像快递员根据地址上的街道名称找到接收方所在的街道，然后再通过门牌号在这条街上找到他的具体位置一样。

互联网中的每个网络的网络号必须是唯一的，同一网络中的主机号也必须是唯一的，不能重复，否则会导致冲突进而影响数据通信。

5. 二进制转换为十进制

我们最熟悉的是十进制数，但是在计算机内部为了表示方便，使用的是二进制数。十进制数在运算的过程中逢十进一，二进制数在运算的过程中逢二进一。

二进制和十进制之间转换的方法有很多，这里只介绍计算效率较高的方法。

以下是二进制转换为十进制的例子，如图 1.19 所示。

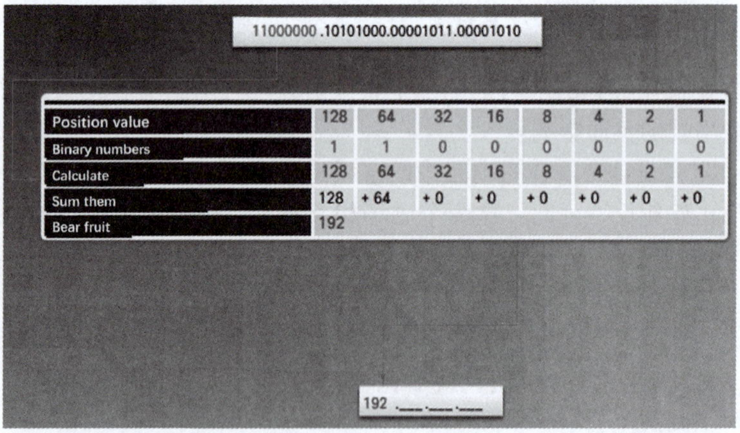

图 1.19　点分十进制计算方法

先写出 8 个二进制位对应的值（见图 1.19 表中第一行，从右边写起，翻倍增加），然后把 IP 地址中的二进制位一一对应上去，把标记为 1 的二进制位（这里是最左边 2 个位）对应的值（这里是 128 和 64）相加，就得到该二进制数对应的十进制的数值。

6. 十进制转换为二进制

十进制数转换为二进制数比较简单的方法叫凑数法。把十进制数 192 转换为二进制数，方法如图 1.20 所示。

图 1.20 将十进制数转换为二进制数

先写出 8 个二进制位对应的值（见图 1.20 表中第一行，从右边写起，翻倍增加），然后用这 8 个数凑够 192，用到的数标记 1，用不到的数标记 0。

如 192=128+64。这里为了凑够 192，用到 128 和 64 这两个数，所以在这两个数下标记 1，其他没有用到的数标记 0，得到二进制数：11000000。

7. IPv4 地址的分类

IPv4 地址分为 A，B，C，D，E 五类。从 IPv4 地址的第 1 个字节识别其属于哪一类。

A 类地址：第 1 个字节从 0 至 127 的 IP 地址，其范围为 0.0.0.0/8 至 127.0.0.0/8，其中 0.0.0.0 和 127.0.0.0 保留不能进行分配。

B 类地址：第 1 个字节从 128 至 191 的 IP 地址，其范围为 128.0.0.0/16 至 191.255.0.0/16。

C 类地址：第 1 个字节从 191 至 223 的 IP 地址，其范围为 192.0.0.0/24 至 223.255.255.0/24。

D 类地址：第 1 个字节从 224 至 239 的 IP 地址。

E 类地址：第 1 个字节从 240 至 255 的 IP 地址。

能在 Internet 上使用的只有 A，B，C 三类。D 类作为组播地址，E 类保留用于实验。

8. 什么是组播地址？除了组播以外还有哪些类型的地址

与网络成功连接的主机可通过单播、广播、组播三种方式的任意一种与其他设备通信，IP 数据包中封装的目的地址分别就是单播地址、广播地址、组播地址。

（1）单播：从一台主机向另一台主机发送数据包的过程。当一个 IP 数据包要发给某一台确定的主机，则这种类型的数据包被称为单播包。这种类型数据包的目的 IP 地址就是那台特定的主机的 IP 地址，这种类型的地址就是单播地址。

（2）广播：从一台主机向该网络中的所有主机发送数据包的过程。当一个 IP 数据包不知道要发给谁，发送的目的主机无法确定时，网络就把这个数据包复制 N 份，发给网络中的每台主机，则这种类型的数据包被称为广播包。这种类型数据包的目的 IP 地址就填写该网络中最大的 IP 地址，这种类型的地址就是广播地址。(想象一下，如果我要发一个快递，可是我只知道他跟我住同一条街，并不知道具体的门牌号，所以我寄快递的时候，在包裹上就写到街道，门牌号写"0"，快递员到了这条街上，就把包裹复制成 N 份，整条街每家每户发一份，这就叫广播。)

（3）组播：从一台主机向选定的一组主机（可能在不同网络中）发送数据包的过程。有时候我们并不是想把数据发给某台特定的主机，也不是发给网络中所有的主机，而是"一些"主机组建成的某个组，则这种类型的数据包被称为组播包。这种类型数据包的目的

IP 地址就填写接收数据的组播组的 IP 地址，这种类型的地址就是组播地址。

9. 公有 IP 地址和私有 IP 地址

公有 IPv4 地址应用于互联网，必须全球唯一。是能在互联网运营商（Internet Service Provider，ISP）路由器之间全面路由的地址，需要申请购买并付费。

私有 IPv4 地址一般应用于组织机构局域网内部。20 世纪 90 年代中期，IPv4 地址空间耗尽。从 A/B/C 三类地址划出了三个网段的专有地址，作为私有 IPv4 地址使用，一定程度上缓解了公有地址枯竭带来的问题。私有 IPv4 地址并不是全球唯一的，任何组织的内部网络都可以免费使用。

具体来说，私有地址块为：
A 类：10.0.0.0/8 或 10.0.0.0 到 10.255.255.255。
B 类：172.16.0.0/12 或 172.16.0.0 到 172.31.255.255。
C 类：192.168.0.0/16 或 192.168.0.0 到 192.168.255.255。

必须知道这些地址块中的地址不能用于互联网，标记私有地址的 IP 数据包会被互联网路由器过滤（丢弃）。

10. Ping 命令

古典名著《三侠五义》中有一个词，叫"投石问路"。原指夜间潜入某处前，先投以石子，看看有无反应，借以探测情况。

Ping 命令的实现过程就类似于"投石问路"。该命令会产生一组数据包，发给目标主机，看是否有回应。如果回应，就说明从发送方到目标主机的整条传输链路是畅通的，而且目标主机存活并给予了应答。通常用于测试目标主机是否存活，或测试从发送方到目标主机的传输链路是否畅通。

11. Ping 命令测试信息分析

本项目中测试得到的结果，对应的主要信息注释如表 1.2 所示。

表 1.2　Ping 命令结果主要信息注释

序号	Ping 命令结果	信息注释
1	Pinging 192.168.0.2 with 32 bytes of data	用 32 字节的数据来 Ping192.168.0.2
2	Reply from 192.168.0.2：bytes = 32 time = 1ms TTL = 128	从 192.168.0.2 主机发来一个应答包：大小 = 32 Byte，数据往返耗时 1 ms，数据包的生命周期 = 128
3	Packets：Sent = 4，Received = 4，Lost = 0 (0% loss)	发送了 4 个包，接收到 4 个应答，丢失 0 个包，丢失率 0%

思考与训练

（1）Internet 的前身是什么？
（2）将你的计算机的 IP 地址转换成二进制的表达方式。

（3）你的计算机所使用的 IP 地址是 A/B/C/D/E 中哪一类的地址？

（4）你的计算机所使用的 IP 地址是私有地址还是公有地址？

（5）查看坐在你旁边的同学所使用计算机的 IP 地址，并用 Ping 命令测试，看你的计算机和他的计算机是否可以正常通信？

（6）EIA/TIA-568A 和 EIA/TIA-568B 标准的线序是怎样的？交叉双绞线和直通双绞线有什么区别？分别用在什么场景？

（7）单模光纤和多模光纤有什么区别？

（8）哪些无线标准用于个人局域网（PAN），并允许设备在 1 到 100 m 的距离内进行通信？

（9）你的手机通过 WiFi 上网是否需要 IP 地址？如果需要，请检查一下你手机的 IP 地址是多少。

Project 1

Understanding Computer Networks

As a starting point in our studies, we will clarify what a computer network is, how it was born and developed. We begin by attempting to create the simplest model of a computer network that can communicate by using a simulator. After completing this project, we should have a relatively intuitive understanding of computer networks and their communication processes, while also boost confidence for subsequent learning.

Learning Objectives

(1) Understand what a computer network is;
(2) Learn about the history of computer network development;
(3) Understand common network topologies in computer networking;
(4) Grasp the definition and role of IP addresses.

Skill Objectives

(1) Learn how to use Cisco Packet Tracer to build simple computer network models;
(2) Master the methods for configuring IP addresses for computers.

Relevant Knowledge

1. What is a Computer Network

The English term for "network" represents a system of interrelated elements. The term can have different meanings in various fields. In our daily life, it often refers to networks such as cable television networks, telephone networks, mobile communication networks, and even interpersonal relationship networks. However, here we mainly discuss communication between computers, i. e., computer networks. What then, is a computer network? The industry generally believes that a collection of interconnected but independent computers can be termed as a computer network system. From this description, we understand that computer networks aim to achieve two goals: the exchange of information between independent, interconnected computers and the sharing of resources among computers.

2. Classification of Computer Networks

Computer networks can be classified into three categories on the scope of coverage:

Local Area Network (LAN): Networks providing access within a small geographical area for users and terminal equipments, typically owned and managed by individuals or IT departments for enterprises, homes, or small businesses.

Metropolitan Area Network (MAN): Networks covering a physical area larger than a LAN but smaller than a WAN, such as a city. MANs are usually operated by a single entity (e.g., a large organization).

Wide Area Network (WAN): Networks providing access across a broad geographical area to other networks, typically owned and managed by telecommunications service providers.

3. Topology of Local Area Networks

The topology of a LAN can be physical or logical, representing different information. Physical topology describes how devices are physically connected with cables, while logical topology describes how devices communicate through physical topology. Physical topology defines the physical interconnection methods of computer terminal systems. In shared-medium LANs, terminal devices can interconnect by using the following physical topologies:

(1) Star: Connecting terminal devices to a central network device. Early star topologies used Ethernet hubs to interconnect terminal devices. Due to the low communication efficiency of hubs, modern star topologies use Ethernet switches. The star topology is easy to install, has good scalability (easy to add or remove terminal devices), and simplifies network troubleshooting, as shown in Figure 1.1.

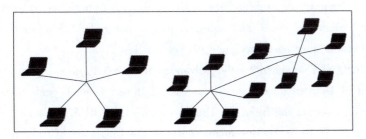

Figure 1.1　Star and extended star topology

(2) Extended Star: The ports on an Ethernet switch are limited and cannot accommodate many users. To expand the network and provide more ports for additional users, switches are interconnected, forming an extended star topology. This extended star is a type of hybrid topology.

(3) Bus: In a bus topology, all terminal systems are connected to each other with a 50-ohm resistor at each end of the bus to absorb electromagnetic signals. Terminal devices are interconnected without infrastructure devices (like hubs or switches). Due to its low cost and easy installation, bus topology using coaxial cables was traditionally used in Ethernet networks.

(4) Ring: In a ring topology, each terminal system connects to its neighbors, forming a ring. Unlike the bus topology, ring topology does not require terminations. Ring topologies were used in traditional Fiber Distributed Data Interface (FDDI) and Token Ring networks. Bus topology and ring topology as shown in Figure 1.2.

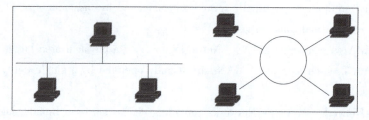

Figure 1.2 Bus topology and ring topology

Task 1: Introduction to Computer Networks—A Brief History of the Internet

Task Objective: Understand the emergence and development history of computer networks.

1. History of the Development of the International Internet

Introduction to Computer Networks

The largest wide area network today is the Internet, which covers the whole world. When we talk about "going online", we refer to accessing the Internet. The development history of the Internet can be traced back to the late 1950s.

In 1950s, communication researchers realized the need to allow regular communication between different computer users and communication networks. This spurred research into decentralized networks, queuing theory, and packet switching.

In 1960, ARPANET, created by the Advanced Research Projects Agency (ARPA) of the U.S. Department of Defense during the Cold War, initiated technological progress and became central to the development of the Internet. The development of ARPANET started with two network nodes, connecting the Network Measurement Center at the University of California, Los Angeles, led by Leonard Kleinrock, and Douglas Engelbart's NLS system at the Stanford Research Institute (SRI) in Menlo Park, California. The third node to join ARPANET was the University of California, Santa Barbara, followed by the University of Utah. By the end of 1971, 15 nodes had been connected to ARPANET.

In June 1973, the Norwegian Seismic Array (NORSAR) became the first network node outside the United States to be connected to ARPANET.

In 1974, Robert Kahn and Vint Cerf proposed the TCP/IP protocol, defining the method of transmitting messages between computer networks (they also received the Turing Award in 2004 for this).

In 1986, the National Science Foundation created the NSFNET, a backbone network based on TCP/IP technology connecting supercomputing centers and academic institutions, with speeds starting from the initial 56 Kbit/s, then T1 (1.5 Mbit/s), and eventually reaching T3 (45 Mbit/s).

Commercial Internet Service Providers (ISPs) emerged in the late 1980s and early 1990s.

ARPANET was decommissioned in 1990.

In the mid to late 1980s, the Internet rapidly expanded in Europe and Australia and extended to Asia in the late 1980s and early 1990s.

In mid-1989, MCI Mail and CompuServe created connections to the Internet and provided E-mail services to 500,000 people.

In March 1990, the first high-speed T1 (1.5 Mbit/s) connection between NSFNET and Europe was established between Cornell University and CERN. Six months later, Tim Berners-Lee wrote the first web browser.

By Christmas of 1990, Tim Berners-Lee had created all the tools needed to run the World Wide Web: Hypertext Transfer Protocol (HTTP), Hypertext Markup Language (HTML), the first web browser, the first web server, and the first website.

In 1995, when NSFNET retired, the Internet was fully commercialized in the United States, lifting the last restrictions on the commercial traffic.

2. History of the Development of the Internet in China

China's Internet is the largest in the world, with the most users and the widest connectivity. However, the overall development time of the Internet in China is relatively short.

On August 25th, 1986, at 4:11 Geneva time and 11:11 Beijing time, the first international E-mail from China was sent by Wu Weimin, then the ALEPH group leader at the Institute of High Energy Physics, to Nobel laureate Jack Steinberger at CERN in Geneva, Switzerland.

In August 1989, the Chinese Academy of Science was responsible for building the "Zhongguancun Education and Research Demonstration Network" (NCFC).

In April 1994, NCFC directly interconnected with the U.S. NSFNET, achieving full functional network connectivity with the Internet, marking the birth of China's earliest international Internet connection.

In 1994, China's first national TCP/IP Internet, the CERNET demonstration network, was completed.

In 1998, CERNET researchers built the first IPv6 testbed in China.

In 2000, China's three major portal websites—Sohu, Sina, and NetEase—were listed on NASDAQ in the United States.

In 2001, the next-generation Internet regional test network was completed and accepted in Beijing.

In 2003, the next-generation Internet demonstration project, CNGI, began implementation.

As of March 2020, China had 904 million Internet users, with 897 million using mobile phones for Internet access, accounting for 99.3% of the total Internet users. Rural Internet users were 255 million, accounting for 28.2% of the overall Internet users. The number of online shopping users reached 710 million, while online payment users were about 768 million.

3. Cyberspace Sovereignty

The Internet has created a new space for human life and, consequently, a new domain for national governance. With the rapid advancement of information technology, the Internet profoundly affects international politics, economics, cultures, society, and military affairs. The mutual promotion of informatization and economic globalization has made the Internet a leading force in innovation-driven development, deeply integrating into every aspect of the society and significantly changing productions and lifestyles, thereby powerfully advancing societal development. The Internet serves as the main artery for the free flow of information, a forum for free exchange of ideas, a community for like-minded individuals, a stronghold for entrepreneurs, wings for market growth, and a fulcrum for national prosperity. The Internet has truly transformed the world into a global village, increasingly making the international community an interdependent community of a shared destiny. Following land, sea, air, and space, cyberspace has become the fifth domain of human production and life.

On July 1st, 2015, the 15th meeting of the 12th Standing Committee of the National People's Congress passed the National Security Law. This new law establishes a network and information security guarantee system and strengthens network management to prevent, stop, and legally punish illegal online activities such as cyber attacks, invasions, espionage, and spreading illegal and harmful information, thereby maintaining national cyberspace sovereignty, security, and development interests. It was the first time that the concept of "cyberspace sovereignty" was explicitly defined.

Cyberspace sovereignty refers to a nation's right to independently decide on the construction, operation, maintenance, and use of networks, as well as the supervision and management of network security. It is a natural extension and manifestation of national sovereignty in cyberspace and an essential part of national sovereignty. As an extension and manifestation of national sovereignty, cyberspace sovereignty embodies the nation's right to independently handle internal and external affairs in cyberspace, enjoying the right to jurisdiction, independence, self-defense, and equality in cyberspace.

4. Cybersecurity Law

With the rapid development of the Internet, various cybersecurity issues have followed: cyber invasions, attacks, and other illegal activities threatening information security; illegal acquisition of citizens' information, infringement of intellectual property, harming citizens' legitimate interests; advocating terrorism and extremism, seriously endangering national security and public interests of society.

The *Cybersecurity Law of the People's Republic of China*, passed on November 7th, 2016, provides legal guarantees for cybersecurity, maintaining cyberspace sovereignty, national security,

Project 1 Understanding Computer Networks

public interests of society, and protecting the legal rights of citizens, legal persons, and other organizations, promoting the healthy development of economic and social informatization.

China's Cybersecurity Law clarifies eight main aspects:

①Elevating the information security level protection system to the level of law.

②Clarifying the security obligations and personal information protection duties of network products and service providers.

③Defining the scope of critical information infrastructures and the main content of its protection system.

④Specifying the overall coordination responsibilities and related supervisory and management duties of national network information departments for cybersecurity work.

⑤Establishing the real-name system on the Internet and the obligation of network operators to provide technical support and assistance to public security and state security organs in maintaining network security and investigating crimes.

⑥Further improving rules for network operators' collection and use of personal information, actively protecting personal network security obligations and responsibilities.

⑦Establishing a national unified monitoring, early warning, information notification, and emergency response system.

⑧Specifying measures to support and promote the development of network security.

The implementation of the Cybersecurity Law has effectively promoted the healthy development of the Internet in China. It has curbed the sale of personal information, cyber fraud, and other behaviors; clarified the "real-name system" online, protected critical information infrastructures, and punished foreign organizations and individuals attacking or sabotaging China's critical information infrastructures; and allowed for "network communication control" during major emergencies.

Classroom Exercise: Share your thoughts on how to comply with China's Cybersecurity Law.

Task 2: Building the Simplest Network Topology—A Two-Node Network

Task Objective: Use Cisco Packet Tracer to build a computer network with two nodes.

We will use Cisco's simulator, Cisco Packet Tracer (version 7.3.x), to build the simplest computer network topology, a network with only two nodes.

Building the Simplest Network Topology

Step 1: Download the latest version of Cisco Packet Tracer from the official website of the Cisco Networking Academy and install it.

Step 2: Run Cisco Packet Tracer, enter the Cisco Networking Academy account and password as required, and enter the workspace, as shown in Figure 1.3.

The account can be created by the class instructor. If you don't have an account, you can log in as a guest by clicking the "Guest Login" button in the lower right corner of the login screen, as shown in Figure 1.4.

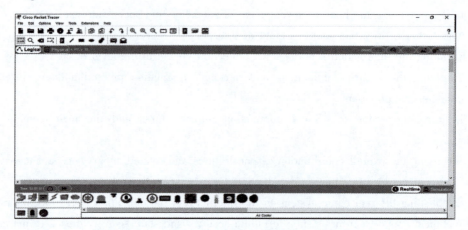

Figure 1.3　Work Interface of Cisco Packet Tracer

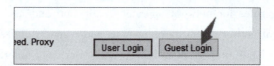

Figure 1.4　Login options

Step 3: In the lower-left corner of the workspace, select the "End Device" icon, choose the first "PC" (Personal Computer) from the devices that appear, and click in the workspace to place a PC, as shown in Figure 1.5.

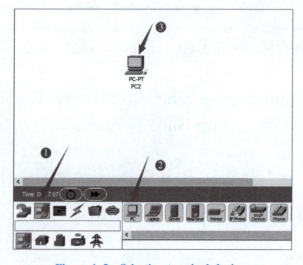

Figure 1.5　Selecting terminal devices

Step 4: Add the second PC using the same method, as shown in Figure 1.6.

Project 1 Understanding Computer Networks

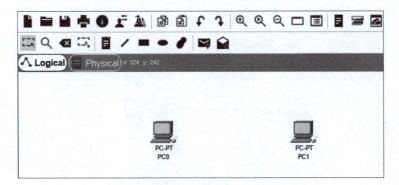

Figure 1.6 Adding another PC

Step 5: Select the Copper Cross-Over media, as shown in Figure 1.7.

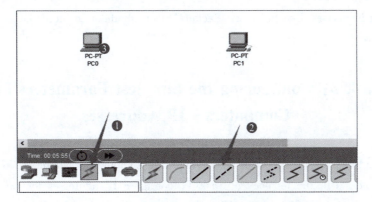

Figure 1.7 Selection of interconnecting media

Step 6: Connect the two PCs using a crossover twisted pair cable. Click on PC0, select the "FastEthernet0" port from the menu that appears, as shown in Figure 1.8.

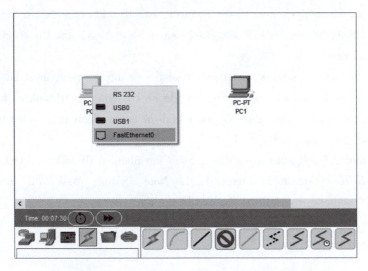

Figure 1.8 Connecting twisted pair to fast ethernet port

Then, connect to the "FastEthernet0" port of PC1 to interconnect the two PCs with the twisted pair cable, as shown in Figure 1.9.

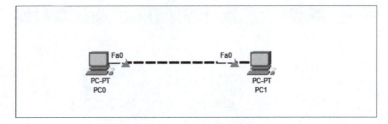

Figure 1.9　Interconnecting Two PC

This forms the simplest computer network topology—two computers directly connected via a crossover twisted pair cable.

Classroom Exercise: Use the Cisco Packet Tracer simulator to build the simplest computer network topology.

Task 3: Configuring the Simplest Parameters for Computers—IP Addresses

Task Objective: Master the method of configuring computer IP addresses in Cisco Packet Tracer (In the topology of Task 1, configure IP addresses for two computers: IP address for PC0 as 192.168.0.1 and for PC1 as 192.168.0.2).

Configuring the Simplest Parameters for Computers— IP Addresses

Like sending a package, data packets transmission in computer network communication needs an address, which is crucial for delivery. In computer network communication, these addresses are called IP addresses. To efficiently transfer data packets from the sender to the receiver, two IP addresses are needed on the IP data packet: the sender's IP address (source IP address) and the receiver's IP address (destination IP address).

Every computer on the network, whether the sender or the receiver, must have an IP address to participate in network communication. So, how do we configure an IP address for a computer?

Step 1: Click on PC0, select the "Desktop" tab in the configuration dialog box, and choose "IP Configuration" as shown in Figure 1.10.

Step 2: In the IP Configuration interface, input the planned IP address: 192.168.0.1 in the "IP Address" field as shown in Figure 1.11. Note: Subnet Mask will auto-generate after configuring the IP address; this will be covered in the next project.

Step 3: Use the same method to configure the IP address for PC1 (192.168.0.2).

Step 4: Communication test.

Project 1 Understanding Computer Networks

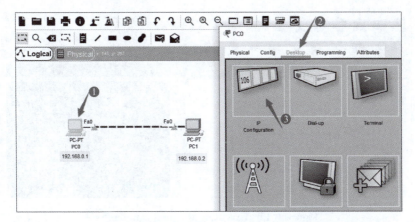

Figure 1.10   Entering the IP Address configuration interface

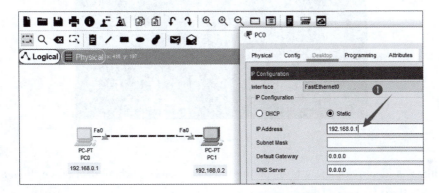

Figure 1.11   Input the planned IP address

Open the Command Prompt tool and enter the CLI interface as shown in Figure 1.12.

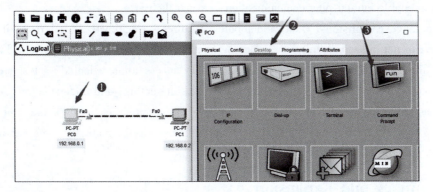

Figure 1.12   Open the Command Prompt tool and enter the CLI interface

Input the test command: ping 192.168.0.2, as shown in Figure 1.13. Press the Enter key to execute the command; if the result shows "Reply from 192.168.0.2: bytes=32 time=1ms TTL= 128," as in Figure 1.14, it indicates successful communication between PC0 and PC1.

27

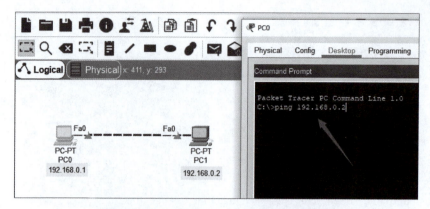

Figure 1.13 Executing the ping command

Figure 1.14 Test results

IP addresses identify different hosts on the Internet, showing where data is sent from and where it is going. Similar to a home address in human society, if you want to send a letter, you need the recipient's address for the postman to deliver it.

Classroom Exercise: Based on the topology from Task 1, configure the IP addresses of two computers to enable mutual ping communication between them.

Competence Expansion: Once a computer acquires a legitimate IP address, it can start participating in network communication, opening another door to the world for people. Essentially, the online life extends our real life. College students access unprecedented vast spaces online, effectively and broadly acquiring information, learning, interacting, and understanding the society. However, as time and energy are limited, excessive online presence reduces the time for other activities. Thus, students should manage their Internet time wisely, regulate online behaviors, and prevent Internet addiction.

Summary and Expansion

1. Elements Constituting a Computer Network

A computer network should consist of the following three elements:
(1) Two or more computers;
(2) Interconnected through some medium;

Project 1 Understanding Computer Networks

(3) Implementing data communication and resource sharing (adhering to common protocols).

2. Common Transmission Media

Besides twisted pair cables, network data can also be transmitted via fiber optics, coaxial cables, and wireless waves.

(1) Twisted Pair.

Twisted Pair is the most commonly-used transmission medium in integrated wiring projects. Twisted pair is divided into Shielded Twisted Pair (STP) and Unshielded Twisted Pair (UTP). Computer networks typically use UTP, consisting of eight insulated copper wires of 22-26 AWG, arranged in four pairs. Each pair is twisted together, hence the name. The twisting cancels out electromagnetic interference from adjacent wires.

The IEEE has defined electrical characteristics for copper cables and categorized UTP wiring based on performance. Cable categories are determined by their capacity to carry higher bandwidth rates. For example, Category 5 cables are usually used for 100BASE-TX Fast Ethernet installations. Higher categories like Enhanced Category 5, Category 6, and 6a have been developed for higher data transfer rates. With the advent of gigabit Ethernet technologies, Category 5e cables are less common, and Category 6 cables are recommended for new installations.

Category 3 cables were initially used for voice communication but later for data transmission. Category 5 and 5e cables support data transmission up to 100 Mbit/s and 1,000 Mbit/s, respectively. Category 6 cables include separators between pairs to support higher speeds, up to 10 Gbit/s. Category 7 cables also support 10 Gbit/s, while Category 8 cables can support up to 40 Gbit/s.

Some manufacturers produce cables that exceed the specifications of TIA/EIA Category 6a and refer to them as Category 7 cables. The ends of UTP cables typically feature RJ45 connectors, as shown in Figure 1.15.

Figure 1.15 Schematic diagram of RJ45 connector

Twisted pair cables are divided into crossover and straight-through types based on the wire order. Crossover cables (EIA/TIA-568A at one end and EIA/TIA-568B at the other) connect similar devices, while straight-through cables (EIA/TIA-568B at both ends) connect different types of devices. The wire order for EIA/TIA-568A and EIA/TIA-568B standards is shown in Table 1.1.

Table 1.1 EIA/TIA-568A and EIA/TIA-568B Wire Sequence Standards

Standard	The color corresponding to the serial number							
	1	2	3	4	5	6	7	8
EIA/TIA-568A	Green and white	Green	Orange and white	Blue	Blue and white	Orange	Brown and white	Brown
EIA/TIA-568B	Orange and white	Orange	Green and white	Blue	Blue and white	Green	Brown and white	Brown
Note: There is a special case where the connection between the computer and the router is made by using a crossover cable.								

(2) Fiber Optics.

Compared to other network media, fiber optics can transmit data over longer distances and at higher bandwidths. Unlike copper cables, fiber optics experience less signal attenuation and are completely immune to electromagnetic and radio frequency interference. Fiber optics are commonly used for interconnecting network devices.

Fiber optics consist of extremely fine, transparent strands of very pure glass, which is about the size of human hair. Data is encoded into light pulses for transmission through fiber optics. They act as waveguides or "light pipes", transmitting light between the two ends with minimal signal loss.

Imagine a hollow paper towel roll covered internally with mirrors, stretching 1,000 meters long, with a small laser emitting Morse code signals at the speed of light. This essentially represents how fiber optics work, albeit with a smaller diameter and using sophisticated light technology.

Fiber optics are generally divided into two types: Single-Mode Fiber (SMF) and Multi-Mode Fiber (MMF).

Single-mode fiber optics contain a very small core and use expensive laser technology to transmit a single beam of light, making them ideal for long-distance transmission over hundreds of kilometers, such as in long-distance telephone and cable TV fibers, as shown in Figure 1.16.

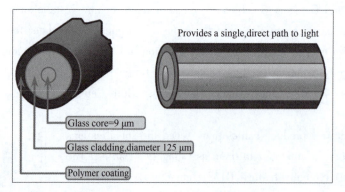

Figure 1.16　Schematic diagram of SMF

Multi-mode fiber optics have a slightly larger core and use LED emitters to send light pulses. This type of fiber, where the LED light enters from different angles, is commonly used in LANs due to support from low-cost LEDs and can provide bandwidths up to 10 Gbit/s over links up to 550 meters, as shown in Figure 1.17.

(3) Radio Waves.

Wireless media uses radio or microwave frequencies to carry electromagnetic signals representing binary digits of data communication. Wireless media offers the best mobility among all media types, with an increasing number of wireless-enabled devices. Wireless is now the primary means for users to connect to home and business networks.

Some limitations of wireless networks include:

①Coverage area: Wireless data communication technologies are well-suited for open environments, but certain building materials and local terrains can limit their effective coverage in buildings.

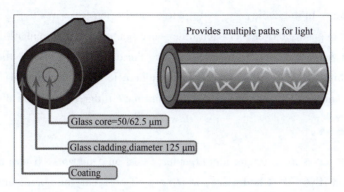

Figure 1.17 Schematic diagram of MMF

②Interference: Wireless communication is susceptible to interference and may be disrupted by common household devices such as cordless phones, certain types of fluorescent lights, microwaves, and other wireless communication devices.

③Security: Wireless communication covers areas without the need for physical media connections. Therefore, devices and users without network access authorization can potentially access the transmission, making network security a crucial component of wireless network management.

④Shared Medium: WLANs operate in half-duplex mode, meaning a device can either send or receive at one time. The wireless medium is shared among all wireless users. Multiple simultaneous accesses to a WLAN can reduce the bandwidth available to each user.

The IEEE and telecommunication industry standards for wireless data communication include the data links and physical layers. Common standardsinclude:

①WiFi(IEEE 802.11)-Known as WiFi, WLAN technology uses a contention protocol called "Carrier-Sense Multiple Access with Collision Avoidance(CSMA/CA)". Wireless NICs must listen before transmitting to ensure the wireless channel is free.

②Bluetooth(IEEE 802.15)-A Wireless Personal Area Network (WPAN) standard, typically known as Bluetooth, communicates by using a device pairing process and covers distances from 1 to 100 meters.

③WiMAX(IEEE 802.16)-Often referred to as WiMAX, this wireless standard adopts a point-to-multipoint topology to provide wireless bandwidth accesses.

④ Zigbee (IEEE 802.15.4)-Zigbee is a specification for low data rate, low power communication, commonly used in industrial and IoT environments like wireless light switches and medical device data collection.

3. Methods for Unshielded Twisted Pair (UTP) Cables to Overcome Electromagnetic and Radio Frequency Interference

UTP cables do not use a shielding layer to counteract electromagnetic and radio frequency interference. Instead, they reduce crosstalk through the following methods:

(1) Cancellation: In the cable, wires are paired, and when two wires are closely aligned, their magnetic fields are opposite and cancel each other out, neutralizing external EMI and RFI signals.

(2) Varying Twists Per Pair: To further enhance the cancellation effect, the number of twists per pair varies within the cable. UTP cables must adhere to specific standards for the number of twists or braids per meter. Different colored wire pairs have different numbers of twists.

4. IP Addresses

An IP address serves as a unique identifier for a host on a network. There are two types of IP addresses, IPv4 and IPv6. IPv4 addresses, which are 32-bit binary numbers, are generally divided into four segments using three dots and converted to decimal notation for readability, as shown in Figure 1.18.

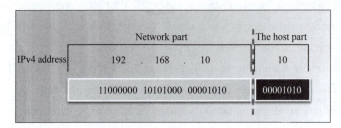

Figure 1.18 Decimal and binary representations of IPv4 addresses

An IP address consists of a network portion and a host portion. The network portion identifies the network number to which the IP address belongs, while the host portion is the specific identifier of the host within that network. The Internet can be linked to a city or a larger area, where a small network is like a street. The network portion is the street number, and the host portion is like the houses' numbers on that street. Network devices identify the target network through the network portion of the IP address and then locate the specific host within that network, similar to how a courier uses the street name to find the street and then the house number to locate the exact address.

Every network in the Internet must have a unique network number, and each host within the same network must also have a unique host number to avoid conflicts and ensure smooth data communication.

5. Binary to Decimal Conversion

We are most familiar with decimal numbers, but binary numbers are used internally in computers for easy representation. In decimal, we count up and add one after reaching ten, whereas in binary, we add one after every two. There are many methods to convert between binary numbers and decimal numbers; here we introduce an efficient method. An example of converting binary numbers to decimal numbers is shown in Figure 1.19.

Project 1 Understanding Computer Networks

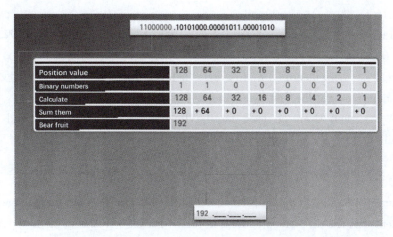

Figure 1.19 Calculation method for decimal point division

To convert an eight-bit binary number to a decimal number, It should be started by writing the corresponding value for each binary digit (starting from the right, doubling each time). Then mapping the binary digits of the IP address to these values. Adding the values corresponding to the binary digits marked as '1' (in this case, the two leftmost digits which are 128 and 64) to get the decimal value.

6. Decimal to Binary Conversion

A simple method for converting decimal numbers to binary ones is called the "making up the number" method. To convert the decimal number 192 into binary one, the procedure should be followed as shown in Figure 1.20.

Figure 1.20 Converting decimal numbers to binary numbers

For example, to convert the decimal number 192 into binary one, list the values for eight binary digits (doubling from right to left) and use these numbers to add up to 192. Mark '1' for numbers used in the sum and '0' for those not used, resulting in the binary number 11000000.

7. Classification of IPv4 addresses

IPv4 addresses are divided into five categories: A, B, C, D, and E, which are identified by the first byte of the address.

Class A addresses have the first bytes between 0 and 127, ranging from 0.0.0.0/8 to 127.0.0.0/8, with 0.0.0.0 and 127.0.0.0 reserved and not allocated.

Class B addresses' first bytes range from 128 to 191, with a range from 128.0.0.0/16 to

191. 255. 0. 0/16.

Class C addresses have the first bytes between 192 and 223, ranging from 192. 0. 0. 0/24 to 223. 255. 255. 0/24.

Class D addresses are identified by first bytes between 224 and 239.

Class E addresses have the first bytes from 240 to 255.

Classes A, B, and C can be used on the Internet. Class D is designated for multicast addresses, while Class E is reserved for experimental purposes.

8. What are Multicast Addresses? What Other Types of Addresses Exist

Hosts connected to a network can communicate with other devices by using unicast, broadcast, or multicast methods. The destination address in the IP packet corresponds to these types: unicast, broadcast, and multicast addresses.

(1) Unicast: The process of sending a packet from one host to another. A packet meant for a specific host is called a unicast packet, and its destination IP is that specific host's IP, known as a unicast address.

(2) Broadcast: Sending a packet from one host to all hosts in the network. When a packet's recipient is unknown and meant for anyone in the network, it's copied and sent to every host, known as a broadcast packet. The destination IP is the network's largest IP, which is the broadcast address.

(3) Multicast: Sending a packet from one host to a selected group of hosts (possibly across different networks). When data isn't meant for a specific host or all hosts but a certain group, it's called a multicast packet. The destination IP is the multicast group's IP address.

9. Public and Private IP Addresses

Public IPv4 addresses are used on the Internet and must be globally unique. They are fully routable between ISP routers and require applications, purchases, and payments. Private IPv4 addresses are typically used within an organization's LAN. To alleviate the exhaustion of public addresses, specific ranges from Classes A, B, and C were designated for private use in the mid-1990s and are not globally unique. They can be freely used internally by any organization.

Specific private address blocks are:

Class A: 10. 0. 0. 0/8 or 10. 0. 0. 0 to 10. 255. 255. 255

Class B: 172. 16. 0. 0/12 or 172. 16. 0. 0 to 172. 31. 255. 255

Class C: 192. 168. 0. 0/16 or 192. 168. 0. 0 to 192. 168. 255. 255

It's important to note that these address blocks cannot be used on the Internet, and IP packets marked with private addresses will be filtered (discarded) by Internet routers.

10. Analysis of Ping Command Test Results

The results obtained in this project, with their corresponding meanings, are outlined in Table 1. 2.

Project 1 Understanding Computer Networks

Table 1.2 Annotations on the Main Information of Ping Command Results

serial number	Ping command results	Main information
1	Pinging 192.168.0.2 with 32 bytes of data	Ping 192.168.0.2 with 32 bytes of data
2	Reply from 192.168.0.2: bytes = 32 time = 1 ms TTL=128	A response packet was sent from the 192.168.0.2 host: size = 32 bytes, data round-trip time of 1 millisecond, and packet lifecycle=128
3	Packets: Sent=4, Received=4, Lost=0(0% loss)	Sent 4 packets, received 4 responses, lost 0 packets, loss rate 0%

Thinking and Training

(1) What was the precursor to the Internet?

(2) Convert your computer's IP address into binary representation.

(3) Which class (A/B/C/D/E) does your computer's IP address belong to?

(4) Is your computer's IP address a private or public address?

(5) Check the IP address of the computer used by the classmate sitting next to you and use the ping command to test if your computer can communicate with theirs.

(6) What are the wire sequences for the EIA/TIA-568A and EIA/TIA-568B standards? What are the differences between crossovers and straight-through twisted pair cables? In what scenarios are each used?

(7) What are the differences between single-mode and multi-mode fiber optics?

(8) Which wireless standards are used for Personal Area Networks (PAN) and allow devices to communicate within a range of 1 to 100 meters?

(9) Does your phone need an IP address to connect to WiFi? If so, check and note your phone's IP address.

项目 2

拓展网络

一个实用的计算机网络肯定不只有两台主机，本项目将计算机网络中的计算机数量由两台拓展到很多台，并解决由此而引发的各种问题。完成本项目的学习后，你就会明白大量的计算机能通过什么网络设备互连？如何确定这些计算机属于相同的网络还是不同的网络？如何实现处于不同网络的计算机之间的相互通信？如何用 WLAN 技术拓展网络？

知识目标

（1）理解交换机和路由器在网络中的作用；
（2）理解子网掩码的意义和作用；
（3）理解网关的意义和作用；
（4）理解交换机和路由器的基本工作原理；
（5）理解虚拟局域网（Virtual Local Area Network，VLAN）的概念和基础知识；
（6）理解无线网络的概念和基础知识。

技能目标

（1）掌握基于命令行的交换机和路由器配置方法；
（2）掌握逻辑网络 IP 地址范围的计算（网络地址、广播地址、可用的主机地址）；
（3）掌握不同逻辑网络间通信的方法；
（4）部署 WLAN，实现移动用户接入网络。

相关知识

项目 1 的任务 1 中的网络拓扑只有两台主机，显然没有太大价值。我们用一根交叉双绞线将两台计算机连接在一起，在这个最简单的网络中，只有两台主机之间能相互访问。

但是我们的办公室一般都是好几个人甚至几十人一起办公的，如何把这几十个员工的计算机连接在一起呢？一般人会直观地想到这样一个全互连型的拓扑，如图 2.1 所示。

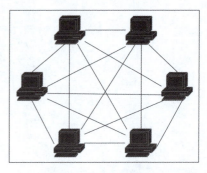

图 2.1　全互连型拓扑

这样的拓扑非常复杂而且成本很高。因为这种联网方式每台计算机需要装好多块网卡才有足够的端口连接到其他的计算机。而且，全世界几亿台计算机，通过这样的方式显然无法实现互连。

项目 1 中介绍的星形拓扑或者拓展星形拓扑结构可以很好地解决这个问题。

拓扑中间有一台网络设备，有很多端口，将旁边几台计算机连接起来。中间那台设备负责终端设备之间的数据转发，叫网络中间设备，旁边的计算机叫网络终端设备。

局域网中负责数据转发的中间设备曾经有过 3 种：集线器、网桥和交换机。集线器每个端口就是一个冲突域，在同一时刻网络中只允许一台主机发送数据，其采用泛洪的方式进行数据转发，速度慢，效率低；后来开发了网桥，很好地隔离了冲突域，构建了 MAC 地址表作为数据转发的依据，但是由于最初的网桥只有两个端口，端口的数量限制了网桥的应用环境。

之后 Kalpana 公司（1994 年被思科收购）发布了第一台以太网交换机。实际上就是把多个网桥集成到一起，相当于是一个提供了多个端口的网桥。现在还习惯把交换机提供互连的方式成为"桥接"，甚至在一些英文的资料中，将以太网交换机称为"Bridge"（桥）的表达依然比较普遍。

目前局域网中最常用的中间设备就是以太网交换机。下面我们就基于 Cisco Packet Tracer 用以太网交换机实现多台计算机的互连。

任务 1：构建小型局域网络——交换机

任务目标：用以太网交换机实现多台计算机的互连。

步骤 1：在 Cisco Packet Tracer 工作区中拖出一台 2960 交换机（Switch），4 台计算机（Laptop），如图 2.2 所示。

构建小型局域网络——交换机

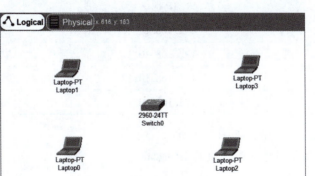

图 2.2 拖出交换机和计算机

步骤 2：用直通双绞线（Copper Straight-Through）将计算机连接到交换机，如图 2.3 所示。

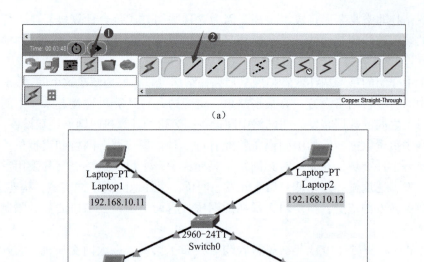

图 2.3 将计算机连接到交换机

(a) 选择直通双绞线；(b) 连接到交换机

注：可以接入交换机的任何一个快速以太网端口（FastEthernet）。

步骤3：按如表2.1所示的IP地址分配表为4台计算机配置IP地址（子网掩码使用默认值）。

表 2.1 IP 地址分配表

主机名	IP 地址	备注
Laptop0	192.168.10.10	默认子网掩码
Laptop1	192.168.10.11	默认子网掩码
Laptop2	192.168.10.12	默认子网掩码
Laptop3	192.168.10.13	默认子网掩码

配置Laptop0的IP地址，如图2.4所示。

用同样的方法配置其他3台主机的IP地址。

步骤4：测试。

从Laptop0主机Ping Laptop1，结果如图2.5所示。

从其他主机相互Ping，结果也应该是通的。

项目 2 拓展网络

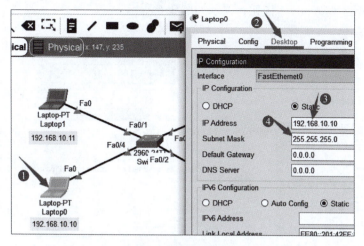

图 2.4 配置 IP 地址

```
C:\>ping 192.168.10.11

Pinging 192.168.10.11 with 32 bytes of data:

Reply from 192.168.10.11: bytes=32 time<1ms TTL=128
Reply from 192.168.10.11: bytes=32 time<1ms TTL=128
Reply from 192.168.10.11: bytes=32 time<1ms TTL=128
Reply from 192.168.10.11: bytes=32 time<1ms TTL=128

Ping statistics for 192.168.10.11:
    Packets: Sent = 4, Received = 4, Lost = 0 (0% loss),
Approximate round trip times in milli-seconds:
    Minimum = 0ms, Maximum = 0ms, Average = 0ms
```

图 2.5 Ping 测试结果

课堂练习：参照本任务，用交换机连接 4 台计算机，配置计算机的 IP 地址，子网掩码用默认值，实现不同计算机之间的相互通信。

素质拓展：小型局域网络的部署为团队协作创造了更好的条件。团队精神已经成为职业发展中最重要的软技能之一，在学习和工作中培养正确的集体观念显得尤为重要。团队协作建立在团队的基础之上，发挥团队精神、互补互助以达到团队最大工作效率。对于团队成员来说，不仅要有个人能力，更需要在不同的位置上各尽所能，与其他成员协调合作。

任务 2：确定计算机所在的网络——子网掩码

任务目标：理解子网掩码并掌握计算一台主机所在网络的网络地址、广播地址和可用的主机地址范围。

在配置 IP 地址的时候，我们用鼠标单击子网掩码（Subnet Mask）文本框，子网掩码就会自动出现。

子网掩码重要吗？我们在 Windows 系统做一个测试：删除网卡的子网掩

确定计算机所在的网络——子网掩码

39

码参数,然后单击"Certain"按钮,提示如图 2.6 所示。

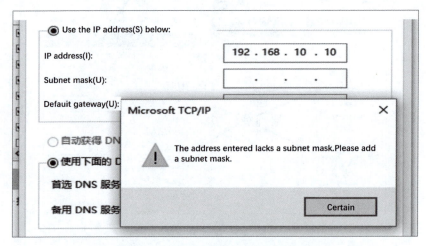

图 2.6 删除子网掩码后的提示信息

出现的警告信息告诉我们,要求必须添加子网掩码,否则无法设置 IP 地址!显然,对于 IP 地址而言子网掩码是必需的。

子网掩码到底是什么?有什么用?为何如此重要?

1. 子网掩码及其作用

子网掩码和 IP 地址本身的表示方法一样,也是采用点分十进制表示法。它是一种用来指明一个 IP 地址的哪些位标志的是主机所在的网络地址以及哪些位标志的是主机地址的位掩码。

子网掩码用于标志 IPv4 地址的网络部分/主机部分,本质上是一个 1 位序列后接 0 位序列的序列。

从以上信息我们归纳出子网掩码的两个特点:

(1) 子网掩码跟 IP 地址一样,也是由 32 位的二进制组成,但是 1 和 0 必须是连续的,不能交叉的,如图 2.7 所示。

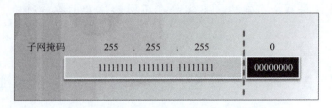

图 2.7 子网掩码的十进制和二进制表示方法

子网掩码还有另外一种很常用的更便捷的表达方式,那就是用二进制的子网掩码中"1"的个数来表示。如图 2.7 所示的子网掩码 255.255.255.0 的二进制表达有 24 个"1",则该子网掩码可以表示成"/24"的方式。因为便捷,所以我们用得最多的其实就是这种方式。

(2) 子网掩码跟 IP 地址一起使用,以便区分在 32 位的 IP 地址中,哪些是网络位部分，哪些是主机位部分。

如图 2.8 所示，子网掩码"1"所对应的，就是 IP 地址的网络位部分，"0"所对应的，就是 IP 地址的主机位部分。

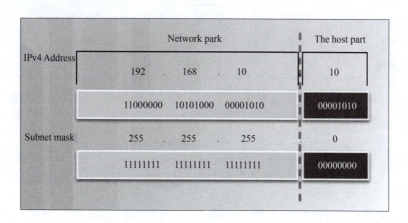

图 2.8　子网掩码和 IP 地址的对应关系

需要注意的是，子网掩码本身并不包含 IP 地址的网络信息，而只是负责告诉主机：你那串 32 位的二进制 IP 地址，哪些位是网络位，哪些位是主机位。如果没有子网掩码，主机就无法知道自己的网络位是哪些，主机位是哪些。而网络位是一个网络的编号，如果无法明确，主机的身份也就不明确。因为不知道自己所处的是哪个网络，导致无法发送数据。类似的，一个人如果不知道自己在哪里是无法发快递的，别人也不知道如何寄快递给他。

2. 计算主机所在网络的网络地址、广播地址和可用的主机地址范围

知道了 IP 地址及其子网掩码，就可以明确主机所在的网络。方法很简单，就是把 IP 地址和子网掩码分别转换成二进制表达，然后进行逻辑"与"（AND）运算。

二进制的逻辑"与"运算如表 2.2 所示。

表 2.2　二进制的逻辑"与"运算

逻辑变量	逻辑运算符	逻辑变量	结果
1	AND	1	1
0	AND	1	0
0	AND	0	0
1	AND	0	0
注：逻辑运算中的 AND 运算，可以用算数运算的"乘法"去类比，其运算结果一样。			

下面我们通过一个案例掌握计算一台主机所在网络的网络地址、广播地址和可用的主机地址范围的方法。

案例：一台主机的 IP 地址为 192.168.10.10，子网掩码为 255.255.255.0，计算出该主机所在网络的网络地址、广播地址和可用的主机地址范围。

步骤1：将十进制的IP地址、子网掩码转换成二进制的表达方式，如表2.3所示。

表2.3　IP地址、子网掩码的十进制和二进制对应关系

IP地址	192	168	10	10
子网掩码	255	255	255	0
二进制IP地址	11000000	10101000	00001010	00001010
二进制子网掩码	11111111	11111111	11111111	00000000

步骤2：将二进制IP地址和二进制子网掩码进行逻辑"与"运算，得到该主机所在网络的网络地址，如表2.4所示。

表2.4　二进制IP地址和二进制子网掩码进行逻辑"与"运算

二进制IP地址	11000000	10101000	00001010	00001010
逻辑运算	AND			
二进制子网掩码	11111111	11111111	11111111	00000000
结果	11000000	10101000	00001010	00000000

所以，IP地址为192.168.10.10/24的主机所在网络的网络地址是：11000000.10101000.00001010.00000000/24。这是一个主机位为全"0"的地址。"0"是二进制数中最小的数字，因此，网络地址就是网络中最小的地址。将网络地址转换成十进制，增加可读性，得到192.168.10.0/24。网络地址是一个网络的编号，是网络在互联网中唯一的标记。

注意：表达一个IP地址的时候，一定要附带子网掩码，否则该IP地址就没有意义。

步骤3：将网络地址主机位部分的每一个二进制位全部换成"1"，就得到该主机所在网络的广播地址，如图2.9所示。

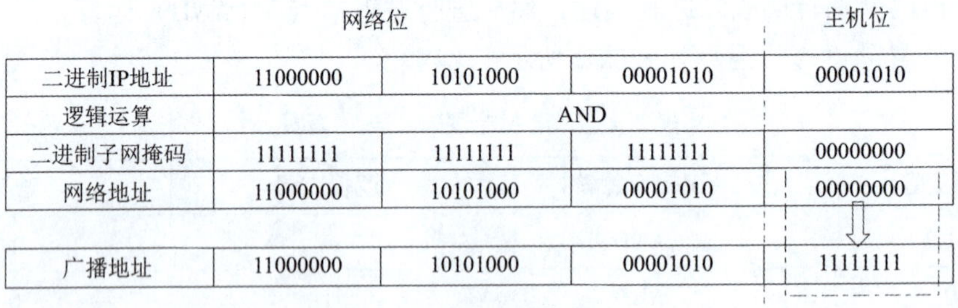

	网络位			主机位
二进制IP地址	11000000	10101000	00001010	00001010
逻辑运算	AND			
二进制子网掩码	11111111	11111111	11111111	00000000
网络地址	11000000	10101000	00001010	00000000
广播地址	11000000	10101000	00001010	11111111

图2.9　广播地址计算过程

广播地址的主机位为全"1"。"1"是二进制数中最大的数字，因此，广播地址就是该网络中最大的地址。将广播地址转换成十进制，增加其可读性，得到192.168.10.255/24。

步骤4：确定可以分配给主机使用的IP地址。

通过以上步骤的计算，得出IP地址为192.168.10.10/24的主机所在网络的网络地址和广播地址分别如表2.5所示。

表 2.5　网络地址和广播地址

地址类型	值	备注
网络地址	192.168.10.0/24	该主机所在网络中的最小 IP 地址
广播地址	192.168.10.255/24	该主机所在网络中的最大 IP 地址

知道了最小 IP 地址和最大 IP 地址，就可以得出一个范围，如图 2.10 所示。

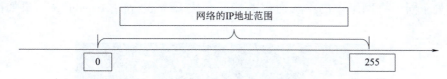

图 2.10　IP 地址坐标区间图

该网络中可用的主机地址就应该为 0~255。但是，192.168.10.0/24 已经被当作网络地址使用，192.168.10.255/24 已经被当作广播地址使用，所以剩下的 192.168.10.1~192.168.10.254 可以被分配给主机使用，这就是该网络中实际可用的主机地址。需要注意的是，192.168.10.0/24 和 192.168.10.255/24 已经分别被网络地址和广播地址占用，不能再被分配给主机使用。

子网掩码用来确定一个 IP 地址的网络位部分和主机位部分，同时确定了该 IP 地址属于哪一个逻辑网络。互联网中的计算机分布在不同的逻辑网络中，但是不管是哪个逻辑网络中的计算机，必须要遵守网络的通信规则才能实现互联网中不同计算机之间的相互通信。

课堂练习：用 ipconfig 命令查看自己上课用的计算机的 IP 地址和子网掩码，并计算自己的计算机所在网络的网络地址、广播地址和可用的主机地址范围。和旁边的同学比较，你们的计算机是否在同一个网络？

素质拓展：计算机网络之所以能实现数据通信，是因为网络中的主机必须遵守共同的规则。有规矩才有方圆，学校和社会也有相应的规则。大学生应当全面了解公共生活领域中的各项法律法规，熟知校纪校规，牢固树立法治观念，以遵纪守法为荣，以违法乱纪为耻，自觉遵守有关的纪律和法律。

任务 3：用路由器连接不同的网络——网关

任务目标：理解路由器的作用及其简单配置，实现不同网络之间的通信。

1. 设计拓扑，并测试不同网络之间主机的通信情况

在本项目任务 1 中的 4 台主机的地址分别为 192.168.10.10/24~192.168.10.13/24，可以计算得出他们的网络地址都一样：192.168.10.0/24。说明这 4 台主机都是在同一个网络中。

设想一下，如果本项目任务 1 中分配的 IP 地址如表 2.6 所示，主机之间的通信还能正常吗？

用路由器
连接不同的
网络——网关

表 2.6　IP 地址分配表及其网络地址

主机名	IP 地址	网络地址	备注
Laptop0	192.168.10.10/24	192.168.10.0/24	同网
Laptop1	192.168.10.11/24	192.168.10.0/24	
Laptop2	192.168.20.12/24	192.168.20.0/24	同网
Laptop3	192.168.20.13/24	192.168.20.0/24	

测试处于同一个网络中的主机通信的情况。从 Laptop0 ping Laptop1，结果如图 2.11 所示。

```
C:\>ping 192.168.10.11

Pinging 192.168.10.11 with 32 bytes of data:

Reply from 192.168.10.11: bytes=32 time=1ms TTL=128
Reply from 192.168.10.11: bytes=32 time<1ms TTL=128
Reply from 192.168.10.11: bytes=32 time=1ms TTL=128
Reply from 192.168.10.11: bytes=32 time<1ms TTL=128

Ping statistics for 192.168.10.11:
    Packets: Sent = 4, Received = 4, Lost = 0 (0% loss),
Approximate round trip times in milli-seconds:
    Minimum = 0ms, Maximum = 1ms, Average = 0ms

C:\>
```

图 2.11　从 Laptop0 ping Laptop1 结果

从 Laptop2 用 ping 命令发送一组 ICMP 数据给 Laptop3（IP 地址：192.168.20.13），结果如图 2.12 所示。

```
C:\>ping 192.168.20.13

Pinging 192.168.20.13 with 32 bytes of data:

Reply from 192.168.20.13: bytes=32 time=1ms TTL=128
Reply from 192.168.20.13: bytes=32 time<1ms TTL=128
Reply from 192.168.20.13: bytes=32 time<1ms TTL=128
Reply from 192.168.20.13: bytes=32 time<1ms TTL=128

Ping statistics for 192.168.20.13:
    Packets: Sent = 4, Received = 4, Lost = 0 (0% loss),
Approximate round trip times in milli-seconds:
    Minimum = 0ms, Maximum = 1ms, Average = 0ms

C:\>
```

图 2.12　从 Laptop2 ping Laptop3 的结果

这两个相同网络里的主机相互通信都没有问题，可是我们测试一下不同网络之间的设备通信（从 Laptop1 ping Laptop3），结果如图 2.13 所示。

```
C:\>ping 192.168.20.13

Pinging 192.168.20.13 with 32 bytes of data:

Request timed out.
Request timed out.
Request timed out.
Request timed out.

Ping statistics for 192.168.20.13:
    Packets: Sent = 4, Received = 0, Lost = 4 (100% loss),
```

图 2.13　从 Laptop1 ping Laptop3 的结果

再从 Laptop0 用 ping 命令发一组 ICMP 数据给 Laptop2（IP 地址：192.168.20.12），结果如图 2.14 所示。

```
C:\>ping 192.168.20.12

Pinging 192.168.20.12 with 32 bytes of data:

Request timed out.
Request timed out.
Request timed out.
Request timed out.

Ping statistics for 192.168.20.12:
    Packets: Sent = 4, Received = 0, Lost = 4 (100% loss),
```

图 2.14 从 Laptop0 ping Laptop2 的结果

结果都是"Request timed out"（请求超时），这说明数据发送出去以后，有可能是接收方 Laptop2 没有收到数据，也有可能是发送方 Laptop1 没有收到接收方 Laptop2 发回的应答包。

发现规律了吗？同网的两台主机之间通信是正常的，不同网络之间的主机无法通信。

结论：交换机只能转发同网络的数据，不能转发不同网络之间的数据（Laptop0 和 Laptop2 处于不同的网络，所以交换机无法转发）。

那么，什么设备能转发不同网络之间的数据呢？答案是：三层网络设备，比如路由器和三层交换机等。

2. 用路由器连接不同的网络

下面我们就将这两个网络通过路由器的快速以太网（Fastethernet，简写为 Fa）端口连接在一起，改造后的拓扑如图 2.15 所示。

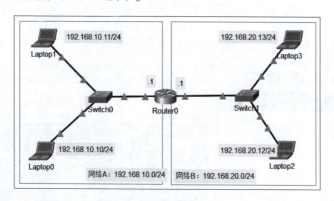

图 2.15 改造后的拓扑

图中路由器 Router0 的 Fa0/0 端口就是网络 A 的网关，负责网络 A 与其他网络的数据流量转发；Fa0/1 端口就是网络 B 的网关，负责网络 B 与其他网络的数据流量转发。

3. 路由器的简单配置

路由器相当于一台计算机，每个网络端口都需要配置一个 IP 地址才可以参与通信。要让路由器正常工作，转发不同网络之间的数据，至少需要配置这几个负责连接不同网络的路由器的端口（网关）。路由器 Router0 端口 IP 地址分配如表 2.7 所示。

表 2.7 路由器 Router0 端口 IP 地址分配

端口名称	IP 地址	备注
Fa0/0	192.168.10.1/24	必须与 Laptop0 和 1 同网
Fa0/1	192.168.20.1/24	必须与 Laptop2 和 3 同网

注意：网关的 IP 地址必须在所连接网络的可用主机地址范围内。（因为网关与主机不同网，则网关就无法接收到主机发来的数据，也就无法将数据转发到其他网络。）

下面开始对路由器进行配置。

步骤 1：用一根 console 线缆将管理计算机 Laptop4 的 RS232 端口连接到路由器的 console 端口，如图 2.16 所示。

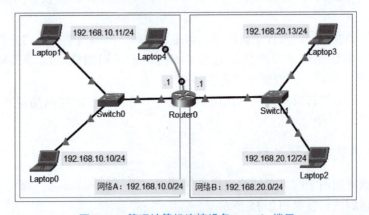

图 2.16 管理计算机连接设备 console 端口

步骤 2：管理计算机通过超级终端软件登录到路由器的控制台界面，如图 2.17 所示。

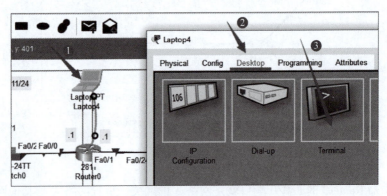

图 2.17 超级终端软件位置

保留连接参数不变，单击"OK"按钮，如图 2.18 所示。

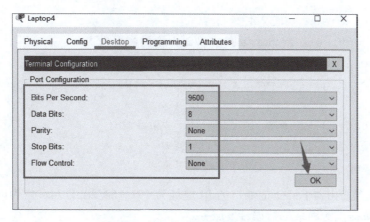

图 2.18 超级终端登录控制台参数

第一次登录路由器的初始界面如图 2.19 示。

图 2.19 第一次登录路由器的初始界面

路由器询问"Would you like to enter the initial configuration dialog?"（您是否希望进入初始配置对话方式？）[yes/no] 表示有两个选项可以输入，这里我们输入"no"，然后回车确定，进入用户模式，如图 2.20 所示。

图 2.20 进入用户模式

步骤 3：配置路由器 Fa0/0 端口的 IP 地址为 192.168.10.1/24，作为网络 A 的网关，为网络 A 的主机提供跨网络的数据转发服务。

```
Router>enable        //进入特权模式
Router#configure terminal    //进入全局配置模式
Router(config)#interface fa0/0    //进入端口 Fa0/0
Router(config-if)#ip address 192.168.10.1 255.255.255.0    //配置端口 IP 地址
Router(config-if)#no shutdown    //激活端口
Router(config-if)#exit    //退出端口
Router(config)#    //回到了全局配置模式
```

步骤 4：配置路由器 Fa0/1 的端口的 IP 地址为 192.168.20.1/24，作为网络 B 的网关，为网络 B 的主机提供跨网络的数据转发服务。

```
Router(config)#interface fa0/1    //进入 Fa0/1 端口
Router(config-if)#ip address 192.168.20.1 255.255.255.0    //配置端口 IP 地址
Router(config-if)#no shutdown    //激活端口
Router(config-if)#exit    //退出端口
Router(config)#    //回到了全局配置模式
```

步骤 5：测试。

先测试主机是否可以 ping 通自己的网关，我们从 Laptop1 ping 路由器的 Fa0/0 端口的地址（192.168.10.1），结果如图 2.21 所示。

```
C:\>ping 192.168.10.1

Pinging 192.168.10.1 with 32 bytes of data:

Reply from 192.168.10.1: bytes=32 time=1ms TTL=255
Reply from 192.168.10.1: bytes=32 time=3ms TTL=255
Reply from 192.168.10.1: bytes=32 time<1ms TTL=255
Reply from 192.168.10.1: bytes=32 time<1ms TTL=255

Ping statistics for 192.168.10.1:
    Packets: Sent = 4, Received = 4, Lost = 0 (0% loss),
Approximate round trip times in milli-seconds:
    Minimum = 0ms, Maximum = 3ms, Average = 1ms
```

图 2.21　Laptop1 ping 路由器的 Fa0/0 端口结果

从 Laptop3 ping 路由器的 Fa0/1 端口的地址（192.168.20.1），结果如图 2.22 所示。

```
C:\>ping 192.168.20.1

Pinging 192.168.20.1 with 32 bytes of data:

Reply from 192.168.20.1: bytes=32 time=1ms TTL=255
Reply from 192.168.20.1: bytes=32 time=1ms TTL=255
Reply from 192.168.20.1: bytes=32 time<1ms TTL=255
Reply from 192.168.20.1: bytes=32 time<1ms TTL=255

Ping statistics for 192.168.20.1:
    Packets: Sent = 4, Received = 4, Lost = 0 (0% loss),
Approximate round trip times in milli-seconds:
    Minimum = 0ms, Maximum = 1ms, Average = 0ms
```

图 2.22　Laptop3 ping 路由器的 Fa0/1 端口的结果

显然，网络 A 和网络 B 的主机都可以跟自己的网关通信，如图 2.23 所示。

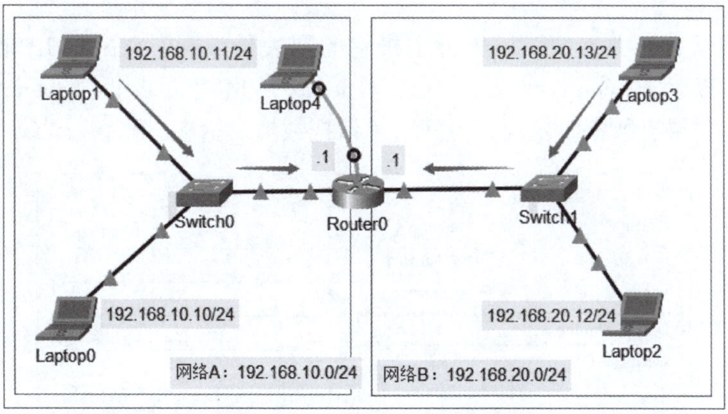

图 2.23 网络内主机和自己网关通信示意图

按已掌握的知识来推演，由于路由器 Router0 在中间可以转发不同网络之间的数据，因此，网络 A 和网络 B 应该可以正常通信。现在我们从 Laptop1 ping Laptop3，测试从网络 A 是否可以传数据到网络 B，结果如图 2.24 所示。

图 2.24 从 Laptop1 ping Laptop3 结果

发现还是"Request timed out"（请求超时）。为什么不同网络之间的 Laptop1 和 Laptop3 还是无法通信呢？

我们以该案例分析不同网络之间通信的过程：Laptop1 把数据封装好以后，准备发出；发出之前要做一个判断，如果接收方是跟自己在一个网络内的，则直接发出；如果接收方是另外一个网络的，则会把数据发给网关，让网关帮忙转发到另外一个网络。用一个生活中的例子做类比：有个人要发一个包裹，如果接收方是跟他同一条街的，则他自己把包裹送去给接收方；如果接收方是另外一条街的，则他会找快递员来帮忙转发这个包裹。

这里是跨网络的通信，显然符合第二种情况：要先把数据发给网关，需要网关帮忙转发到另外一个网络。可问题是，从来没人告诉 Laptop1 它的网关是谁，IP 地址是多少！（如上类比：如果发件人需要把快递发到另外一条街，但他又不知道快递员是谁，显然无法将快递发出去。）

所以在这里，解决问题的办法是要告诉网络内所有的主机，它们的网关是谁，即网关的 IP 是什么。

4. 为主机配置网关参数

如何给主机配置网关参数，以告诉主机它的网关是谁呢？在配置 IP 地址的界面，如图 2.25 所示可以为 Laptop1 配置网关（Gateway）信息，告诉 Laptop1 它的网关是 192.168.10.1，如果有数据要发到其他网络，请把数据发给它，让它帮忙转发。

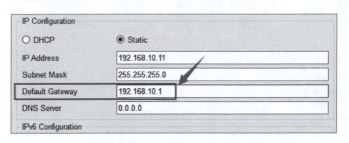

图 2.25　为 Laptop1 配置网关（Gateway）信息

用同样的方法配置其他主机的网关。注意：网络 B 的网关是 192.168.20.1。

最后，我们从 Laptop1 ping Laptop3，测试一下不同网络之间的通信是否正常，结果如图 2.26 所示。

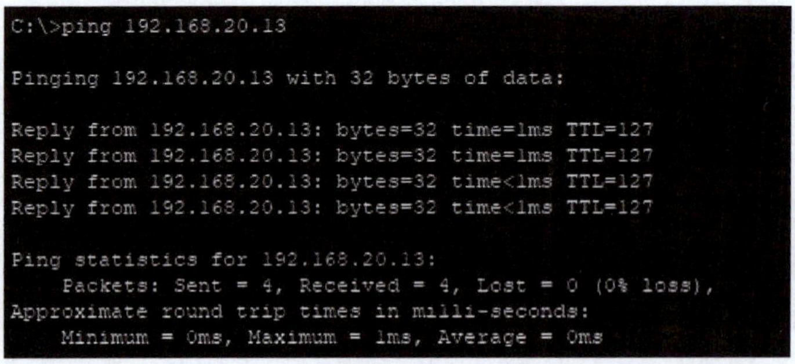

图 2.26　从 Laptop1 ping Laptop3 结果

Laptop1 和 Laptop3 能正常通信，说明成功实现了跨网络的数据传输。

课堂练习：在任务 2 拓扑的基础上，拓展两个网络。参照本任务，用路由器连接两个不同的网络，适当配置，实现全网用户可以相互 ping 通。

素质拓展：网关（Gateway）是不同网络之间的连接点，类似于国家和国家之间的口岸。网关承载着不同网络之间的数据通信，为了过滤危险数据，我们通常会在网关部署安全策略，以确保不同网络之间传输数据的安全性。我国接入 Internet 的节点更是承担着维护网络空间安全、国家主权和国家安全的重要任务。我们坚持走和平发展道路，既重视自身安全，又重视共同安全，打造人类命运共同体，推动世界朝着互利互惠、共同安全的目标相向而行。

任务 4：用三层交换机连接不同的网络——SVI（VLAN、Trunk）

任务目标：用三层交换机实现不同网络之间的通信，并理解其作用。

路由器可以完成不同网络之间的数据转发，一般情况下每个网络会占用路由器的一个端口。如果某个组织的网络划分了很多个逻辑网络，那么如果用路由器来连接不同的网络，就需要大量的端口。而路由器的端口数量是很少的，比如 Cisco 2811 路由器的默认端口数量只有两个。如图 2.27 所示是 Cisco 2811 路由器背板界面。

用三层交换机连接不同的网络——SVI（VLAN、Trunk）

图 2.27　Cisco 2811 路由器背板界面

当然，我们可以采购更多的模块插入扩展槽，以提供更多的网络端口。但是这样的方式无疑增加了网络部署的成本，显然不应该成为首选。

比较普遍的做法是：用三层交换机替代路由器，负责不同网络之间的数据转发。

1. 用三层交换机连接不同的网络

把本项目任务 3 中的路由器换成三层交换机 Cisco 3560 以后，拓扑如图 2.28 所示。

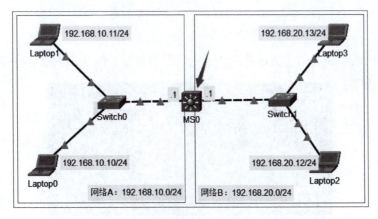

图 2.28　部署 Cisco 3560 三层交换机

这里三层交换机 MS0 的 Fa0/1 端口可以当作网络 A 的网关，Fa0/2 端口可以当作网络 B 的网关。

2. 配置三层交换机，实现不同网络之间的主机通信

路由器 Router0 端口 IP 地址分配如表 2.8 所示。

表 2.8　路由器 Router0 端口 IP 地址分配

端口名称	IP 地址	备注
Fa0/1	192.168.10.1/24	必须与 Laptop0 和 1 同网
Fa0/2	192.168.20.1/24	必须与 Laptop2 和 3 同网

同样的方法，连接 MS0 的 console 端口，通过超级终端 Terminal 工具登录控制台界面，配置如下：

```
Switch>enable    //进入特权模式
Switch#configure terminal     //进入全局配置模式
Switch(config)# hostname MS0     //将交换机的名字设置为“MS0”(不是必需的,对通信无影响,只是方便管理,但这是一个好习惯)
MS0 (config)#interface fa0/1     //进入 Fa0/1 端口
MS0 (config-if)#no switchport     //关闭端口的交换特性(默认情况交换机端口不允许配置 IP 地址)
MS0 (config-if)#ip address 192.168.10.1 255.255.255.0     //配置 IP 地址作为网络 A 的网关
MS0 (config-if)#exit     //退出 Fa0/1 端口
MS0 (config)#interface fa0/2     //进入 Fa0/2 端口
MS0 (config-if)#no switchport     //关闭端口的交换特性
MS0 (config-if)#ip address 192.168.20.1 255.255.255.0     //配置 IP 地址作为网络 B 的网关
MS0 (config-if)#exit     //退出 Fa0/2 端口
MS0 (config)#ip routing     //启用 IP 路由功能(Cisco 3560 三层交换机具备路由功能,但是默认是关闭的,需要手动启用)
```

从 Laptop1 ping Laptop3，测试跨网络的通信是否正常，结果如图 2.29 所示。

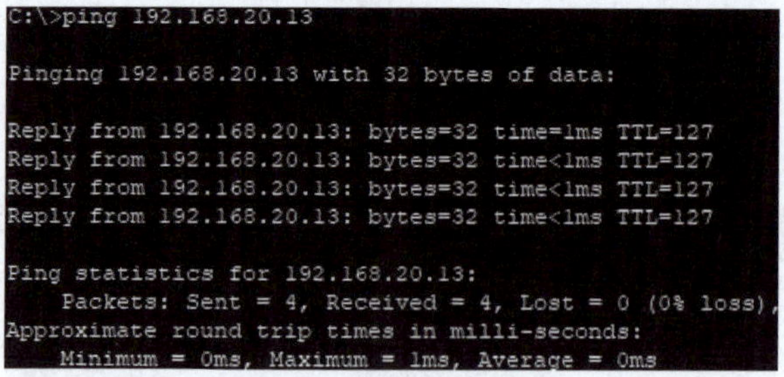

```
C:\>ping 192.168.20.13

Pinging 192.168.20.13 with 32 bytes of data:

Reply from 192.168.20.13: bytes=32 time=1ms TTL=127
Reply from 192.168.20.13: bytes=32 time<1ms TTL=127
Reply from 192.168.20.13: bytes=32 time<1ms TTL=127
Reply from 192.168.20.13: bytes=32 time<1ms TTL=127

Ping statistics for 192.168.20.13:
    Packets: Sent = 4, Received = 4, Lost = 0 (0% loss),
Approximate round trip times in milli-seconds:
    Minimum = 0ms, Maximum = 1ms, Average = 0ms
```

图 2.29　从 Laptop1 ping Laptop3 结果

测试结果显示通信正常。

但是，关闭交换机端口交换功能，将交换机端口变成三层路由端口的实现方法并不适合通用的环境。

下面举一个例子来说明这个问题：我们把网络 A 和 B 具体化，假设网络 A 和网络 B 是一个公司两个部门（销售部和研发部）的逻辑网络。销售部人员不断增加，网络工程师发现销售部网络的交换机的端口已经完全被占用，销售部后面新来的员工已经没有端口可用。要将销售部的新员工的计算机接入到其他部门的交换机，以便访问网络，如图 2.30 所示。

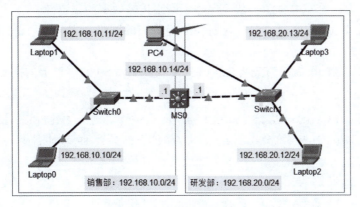

图 2.30 增加一台计算机 PC4

但是 PC4 要和自己同部门的其他计算机通信时，数据要经过 MS0 的 Fa0/2。但是这个端口是研发部的网关，与 PC4 处于不同的网络，注定无法通信，因此导致 PC4 和自己部门的计算机无法通信，所以这样的设计显然是有局限的。

有没有什么方法能对网络做进一步的优化，以提高其可扩展性呢？当然有，用交换机虚拟端口（Switch Virtual Interface，SVI）技术可以解决这个问题。

3. 用 SVI 技术优化网络

解决的方法是引进 VLAN 的概念。将一个物理网络划分成多个逻辑网络，以逻辑网络为单位进行管理。

逻辑网络是相对于物理网络的一个概念。物理网络是通过物理介质连接成一体的网络系统；而逻辑网络是在物理网络基础上的分组，是可以根据需求进行划分的集合。一个物理网络可以划分不同的逻辑网络，而物理网络本身是不变的。通常我们会根据地理位置或用户群的不同来划分逻辑网络。

VLAN 是一组逻辑上的设备和用户，这些设备和用户并不受物理位置的限制，可以根据功能、部门及应用等因素将它们组织起来，相互之间的通信就好像它们在同一个网络中一样。

在我们的这个案例中，整个网络是一个物理网络，所有设备都是通过设备和介质连接在一起的。但是这里我们根据部门规划成了两个网络地址不一样的网络，也就是两个逻辑网络，那么这两个逻辑网络我们可以称为是两个 VLAN。

将不同部门的网络规划成两个不同的 VLAN，主要目的是为了区分与隔离不同部门之间的数据流量。首先，划分了 VLAN 的网络交换机，在接收到用户发来的数据以后，会在数据帧上打一个标签（tag），标志其作为特定 VLAN 的数据，以区分不同逻辑网络的流量。其次，二层交换机不会转发不同 VLAN 之间的数据，而是将用户数据限制在自己的逻辑网络内传输，从而避免了某些逻辑网络的用户非法获取其他网络用户的数据，导致安全事件的发生。

因为可以隔离不同逻辑网络之间的流量，VLAN 还用来隔离局域网的广播数据，限制了广播包在某个特定 VLAN 中传播，保护了其他 VLAN 的用户免受大规模广播包的侵袭而消耗大量 CPU、内存、带宽等资源，从而保障了网络传输效率的稳定性。

下面我们用 VLAN 的概念，重新规划和配置我们的网络，实现两个部门主机之间的相互通信。

为了不影响后续的配置，首先将之前在三层交换机 MS0 的端口上的配置全部清空，命令如下：

```
MS0(config)# interface fastEthernet 0/1    //进入 FastEthernet 0/1 端口
MS0(config-if)#no ip address    //将原先配置的 IP 地址删除
MS0(config-if)#switchport    //恢复端口的交换功能
MS0(config-if)#exit    //退出端口
MS0(config)# interface fastEthernet 0/2    //进入 FastEthernet 0/2 端口
MS0(config-if)#no ip address    //将原先配置的 IP 地址删除
MS0(config-if)#switchport    //恢复端口的交换功能
MS0(config-if)#exit    //退出端口
```

然后开始进行本任务的实施。

步骤 1：为两个部门的用户规划 VLAN，如表 2.9 所示。

表 2.9　VLAN 规划

部门名称	VLAN 编号	VLAN 名称	网关地址	交换机端口
销售部	10	Sales	192.168.10.1/24	Switch0 的 Fa0/1，Fa0/4；Switch1 的 Fa0/1
研发部	20	Development	192.168.20.1/24	Switch1 的 Fa0/12，Fa0/3

步骤 2：按规划表，为销售部创建 VLAN 10，并将销售部用户计算机所接入的交换机端口划入的 VLAN 10。

在 Switch0 上创建 VLAN 10，并把销售部两台计算机占用的端口 Fa0/1，Fa0/4 划入 VLAN 10，命令如下：

```
Switch>
Switch>enable    //进入特权模式
Switch#configure terminal    //进入全局配置模式
Switch(config)#hostname Switch0    //修改主机名(方便管理)
Switch0(config)#vlan 10    //创建 VLAN 10
Switch0(config-vlan)#name Sales    //为 VLAN 10 定义一个名字:Sales
Switch0(config-vlan)#exit    //退出 VLAN 10 的配置
Switch0(config)#interface fa0/1    //进入 Fa0/1 端口
Switch0(config-if)#switchport mode access    //将端口的工作模式改为用户访问模式
Switch0(config-if)#switchport access vlan 10    //将端口划入 VLAN 10
```

```
    Switch0(config-if)#exit
    Switch0(config)#interface fa0/4        //进入 Fa0/4 端口
    Switch0(config-if)#switchport mode access        //将端口的工作模式改为
用户访问模式
    Switch0(config-if)#switchport access vlan 10        //将端口划入 VLAN 10
    Switch0(config-if)#exit        //退出 Fa0/4 端口
```

在 Switch1 上创建 VLAN 10，并把销售部计算机占用的端口 Fa0/1 划入 VLAN 10，命令如下：

```
    Switch>
    Switch>enable        //进入特权模式
    Switch#configure terminal        //进入全局配置模式
    Switch(config)#hostname Switch1        //修改主机名(方便管理)
    Switch1(config)#vlan 10        //创建 VLAN 10
    Switch1(config-vlan)#name Sales        //为 VLAN 10 定义一个名字:Sales
    Switch1(config-vlan)#exit        //退出 VLAN 10 的配置
    Switch1(config)#interface fa0/1        //进入 Fa0/1 端口
    Switch1(config-if)#switchport mode access        //将端口的工作模式改为
用户访问模式
    Switch1(config-if)#switchport access vlan 10        //将端口划入 VLAN 10
    Switch1(config-if)#exit        //退出 Fa0/1 端口
    Switch1(config)#
```

步骤3：按规划表，为研发部创建 VLAN 20，并将销售部用户计算机所接入的交换机端口划入的 VLAN 20。

在 Switch1 上创建 VLAN 20，并把销售部计算机占用的端口 Fa0/12 和 Fa0/13 划入 VLAN 20，命令如下：

```
    Switch1(config)#vlan 20        //创建 VLAN 20
    Switch1(config-vlan)#name Development        //为 VLAN 10 定义一个名
字:Development
    Switch1(config-vlan)#exit        //退出 VLAN 20 的配置
    Switch1(config)#interface range fa0/12-13        //进入 Fa0/12,Fa0/13 端
口群
    Switch1(config-if)#switchport mode access        //将端口的工作模式改为
用户访问模式
    Switch1(config-if)#switchport access vlan 20        //将端口划入 VLAN 20
    Switch1(config-if)#exit        //退出 Fa0/1 端口
    Switch1(config)#
```

检查 VLAN 的配置：

在 Switch0 查看 VLAN 数据库信息，如图 2.31 所示。

```
Switch0#show vlan

VLAN Name                             Status    Ports
---- -------------------------------- --------- -------------------------------
1    default                          active    Fa0/2, Fa0/3, Fa0/5, Fa0/6
                                                Fa0/7, Fa0/8, Fa0/9, Fa0/10
                                                Fa0/11, Fa0/12, Fa0/13, Fa0/14
                                                Fa0/15, Fa0/16, Fa0/17, Fa0/18
                                                Fa0/19, Fa0/20, Fa0/21, Fa0/22
                                                Fa0/23, Gig0/1, Gig0/2
10   Sales                            active    Fa0/1, Fa0/4
1002 fddi-default                     active
1003 token-ring-default               active
1004 fddinet-default                  active
1005 trnet-default                    active
```

图 2.31　Switch0 的 VLAN 数据库信息

在 Switch1 查看 VLAN 数据库信息，如图 2.32 所示。

```
Switch1#show vlan

VLAN Name                             Status    Ports
---- -------------------------------- --------- -------------------------------
1    default                          active    Fa0/2, Fa0/3, Fa0/4, Fa0/5
                                                Fa0/6, Fa0/7, Fa0/8, Fa0/9
                                                Fa0/10, Fa0/11, Fa0/14, Fa0/15
                                                Fa0/16, Fa0/17, Fa0/18, Fa0/19
                                                Fa0/20, Fa0/21, Fa0/22, Fa0/23
                                                Gig0/1, Gig0/2
10   Sales                            active    Fa0/1
20   Development                      active    Fa0/12, Fa0/13
1002 fddi-default                     active
1003 token-ring-default               active
1004 fddinet-default                  active
1005 trnet-default                    active
```

图 2.32　Switch1 的 VLAN 数据库信息

检查发现所有 VLAN 10 和 VLAN 20 已经创建，并已经把用户所占用端口划入对应的 VLAN 中。

步骤 4：因为交换机之间的链路要承载所有 VLAN 的数据，不属于某个具体 VLAN，不是为用户提供接入服务的，所以不能将交换机之间互连的端口设置为 access 模式，而需要将交换机 Switch0 和交换机 MS0 之间的链路设置为干道（Trunk）。

将交换机 Switch0 的 Fa0/24 端口和交换机 MS0 的 Fa0/1 端口的交换模式设置为干道（Trunk），命令如下：

先配置二层交换机 Switch0。

```
    Switch0(config)# interface fastEthernet 0/24    //进入 FastEthernet
0/24 端口
    Switch0(config-if)#switchport mode trunk        //将端口工作模式设置为
Trunk(干道)模式
    Switch0(config-if)#exit     //退出端口
    Switch0(config)#
```

配置核心交换机 MS0：

```
MS0(config)# interface fastEthernet 0/1    //进入 FastEthernet 0/1 端口
MS0(config-if)#switchport trunk encapsulation dot1q    //定义封装协议为 dot1q
MS0(config-if)#switchport mode trunk    //将端口工作模式设置为 Trunk(干道)模式
MS0(config-if)#exit    //退出端口
MS0(config)#
```

步骤 5：将交换机 Switch1 和交换机 MS0 之间的链路设置为干道（Trunk）。

将交换机 Switch1 的 Fa0/24 端口和交换机 MS0 的 Fa0/2 端口的交换模式设置为干道（Trunk），命令如下：

配置二层交换机 Switch1：

```
Switch1(config)# interface fastEthernet 0/24    //进入 FastEthernet 0/24 端口
Switch1(config-if)#switchport mode trunk    //将端口工作模式设置为 Trunk(干道)模式
Switch1(config-if)#exit    //退出端口
Switch1(config)#
```

配置核心交换机 MS0：

```
MS0(config)# interface fastEthernet 0/2    //进入 FastEthernet 0/2 端口
MS0(config-if)#switchport trunk encapsulation dot1q    //定义封装协议为 dot1q
MS0(config-if)#switchport mode trunk    //将端口工作模式设置为 Trunk(干道)模式
MS0(config-if)#exit    //退出端口
MS0(config)#
```

检查交换机 Swith0 的 Trunk 端口状态，如图 2.33 所示。

```
Switch0#show interfaces trunk
Port       Mode        Encapsulation    Status         Native vlan
Fa0/24     on          802.1q           trunking       1
```

图 2.33 Swith0 的 Trunk 端口状态

看到交换机互连端口 Fa0/24 的模式（Mode）是"on"，状态（Status）为"trunking"，则表示端口状态是正常的。

检查其他交换机的 Trunk 端口状态，如图 2.34 和图 2.35 所示。

```
Switch1#show interfaces trunk
Port       Mode        Encapsulation    Status         Native vlan
Fa0/24     on          802.1q           trunking       1
```

图 2.34 Swith1 的 Trunk 端口状态

```
MS0#show interfaces trunk
Port       Mode       Encapsulation    Status       Native vlan
Fa0/1      on         802.1q           trunking     1
Fa0/2      on         802.1q           trunking     1
```

图 2.35　MS0 的 Trunk 端口状态

以上信息说明，交换机 Switch1 和 MS0 所有 Trunk 端口的状态正常。

步骤 6：在 MS0 上创建 VLAN 10 和 VLAN 20，以承载 VLAN 10 和 VLAN 20 的数据。

MS0(config)#vlan 10　　//创建 VLAN 10
MS0(config-vlan)#name Sales　　//为 VLAN 10 起一个名字:Sales
MS0(config-vlan)#exit　　//退出 VLAN 10 的配置
MS0(config)#vlan 20　　//创建 VLAN 20
MS0(config-vlan)#name Development　　//为 VLAN 20 起一个名字:Development
MS0(config-vlan)#exit　　//退出 VLAN 20 的配置

检查 MS0 上的 VLAN 数据库信息，结果如图 2.36 所示。

```
MS0#show vlan

VLAN Name                       Status    Ports
---- ---------------------      --------- -------------------------------
1    default                    active    Fa0/3, Fa0/4, Fa0/5, Fa0/6
                                          Fa0/7, Fa0/8, Fa0/9, Fa0/10
                                          Fa0/11, Fa0/12, Fa0/13, Fa0/14
                                          Fa0/15, Fa0/16, Fa0/17, Fa0/18
                                          Fa0/19, Fa0/20, Fa0/21, Fa0/22
                                          Fa0/23, Fa0/24, Gig0/1, Gig0/2
10   Sales                      active
20   Development                active
1002 fddi-default               active
1003 token-ring-default         active
1004 fddinet-default            active
1005 trnet-default              active
```

图 2.36　MS0 的 VLAN 数据库信息

已经成功创建 VLAN 10 和 VLAN 20，而且 VLAN 的名字也是一致的。（注意：VLAN 的名字在各交换机上必须一致，否则网络设备不认为是同一个 VLAN，哪怕 VLAN 号是一样的。）

此时，PC4 应该可以跨交换机与同部门的用户通信，测试结果如图 2.37 所示。

```
C:\>ping 192.168.10.11

Pinging 192.168.10.11 with 32 bytes of data:

Reply from 192.168.10.11: bytes=32 time=1ms TTL=128
Reply from 192.168.10.11: bytes=32 time<1ms TTL=128
Reply from 192.168.10.11: bytes=32 time<1ms TTL=128
Reply from 192.168.10.11: bytes=32 time<1ms TTL=128

Ping statistics for 192.168.10.11:
    Packets: Sent = 4, Received = 4, Lost = 0 (0% loss),
Approximate round trip times in milli-seconds:
    Minimum = 0ms, Maximum = 1ms, Average = 0ms
```

图 2.37　PC4 ping Laptop1 结果

步骤 7：在 MS0 上配置 VLAN 10 和 VLAN 20 端口的 IP 地址，分别为 VLAN 10 和 VLAN 20 的用户提供网关服务，实现销售部和研发部的数据转发。

在 MS0 中命令如下：

```
MS0(config)#interface vlan 10          //进入 VLAN 10 端口(注意,是端口)
MS0(config-if)#ip address 192.168.10.1 255.255.255.0    //配置 IP 地址,用作 VLAN 10 的网关
MS0(config-if)#exit          //退出
MS0(config)#interface vlan 20          //进入 VLAN 20 端口(注意,是端口)
MS0(config-if)#ip address 192.168.20.1 255.255.255.0    //配置 IP 地址,用作 VLAN 20 的网关
MS0(config-if)#exit          //退出
MS0(config)#
```

销售部和研发部的用户分别 ping 自己的网关，应该可以正常通信，结果如图 2.38 和 2.39 所示。

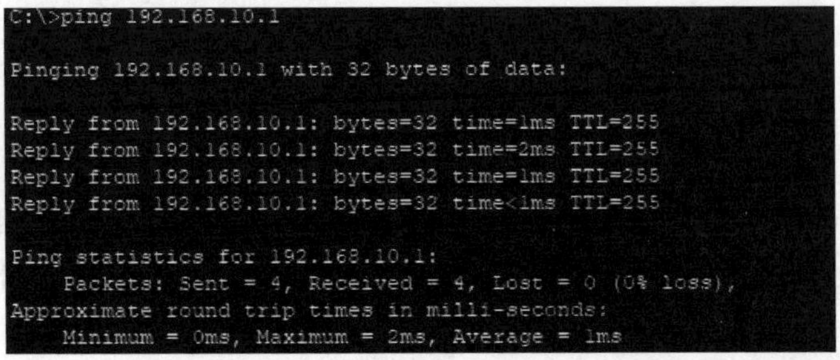

图 2.38　Laptop1 ping 自己网关的结果

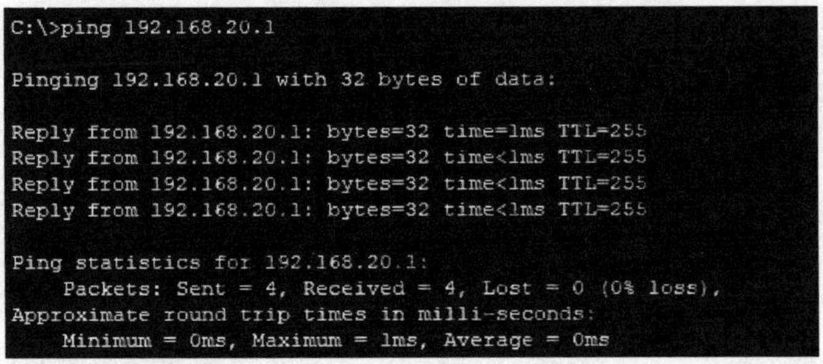

图 2.39　Laptop3 ping 自己网关的结果

步骤 8：测试不同网络之间的通信。

从销售部的 Laptop1 ping 研发部的 Laptop3，结果如图 2.40 所示。

本任务涉及的 VLAN、Trunk 和 SVI 技术虽然基础，但因为几乎每个网络都会用到，所以应该是本书中最实用、最关键的技术之一。掌握了这些技术，对后面学习会有很大的帮助。

课堂练习：基于本项目任务 3 的拓扑，用三层交换机替换路由器。在三层交换机用 SVI 作为用户的网关，适当配置，实现全网互通。

图 2.40　Laptop1 ping Laptop3 的结果

素质拓展：SVI 和 VLAN 是一种虚拟化的思路，解决了网络的可扩展性问题。当传统技术遇到瓶颈的时候，需要创造性地提出新的解决问题的办法。我们在成长过程中也会遇到各种问题，要学会积极应对，不断增强以改革创新推动技术进步，在改革创新中奉献服务社会、实现人生价值的崇高责任感和使命感，以时不我待、只争朝夕的紧迫感投身改革创新的实践中。

任务 5：部署无线局域网——WLAN

部署无线局域网——WLAN

任务目标：掌握基础的 WLAN 技术，为移动用户提供网络访问服务。

有一个段子：几个国家的考古学家组成一个小组，协作完成一个项目。工作间隙大家闲聊。埃及的考古学家说他们在考古的时候，在金字塔里发现了同轴电缆，说明那个时候他们的祖先可能已经用上电缆来传输数据了；英国的考古学家不甘示弱，说在他们的古城堡下挖出的是光纤，说明他们很早以前就已经开始使用光纤来通信了；中国的考古学家不慌不忙说：我们国家什么都没有挖到，可能我们的祖先一直都是用无线来传输数据的。

无线传输应该是人类通信史上的一个里程碑。它不使用任何导线或传输电缆连接的局域网，而使用无线电波或电场与磁场作为数据传送的介质，为用户的接入提供了很大的便利。

最简单的 WLAN 组网方式，就是部署一台无线接入点（Access Point，AP），用户计算机通过无线网卡接入 AP。Cisco Packet Tracer 的无线 AP 位置如图 2.41 所示。

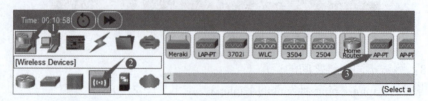

图 2.41　Cisco Packet Tracer 的无线 AP 位置

下面我们通过一个案例来一起学习无线局域网的部署。销售部的人越来越多，原来交换机的端口已经不够用，网络管理员计划部署一台 AP 让销售部员工通过无线方式接入网络。

步骤1：在网络中部署无线AP，无线AP拓扑如图2.42所示。

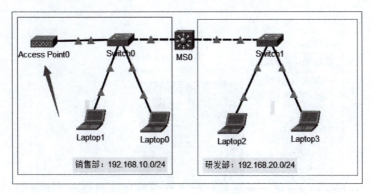

图 2.42　无线 AP 拓扑

步骤2：给 PC4 装上无线网卡，让其通过无线 AP 访问网络。
先删除 PC4 连接的双绞线，关闭 PC4 的电源，主机电源位置如图 2.43 所示。

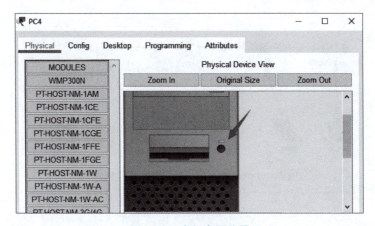

图 2.43　主机电源位置

把网卡拖动到"MODULES"，卸下原来的 RJ45 网卡，主机网卡位置如图 2.44 所示。

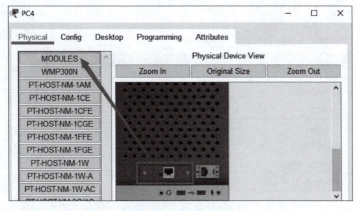

图 2.44　主机网卡位置

在 MODULES 中将型号为"WMP300N"的无线网卡拖动到网卡插槽，如图 2.45 所示。

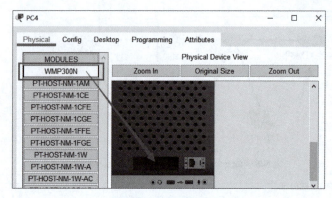

图 2.45　安装无线网卡 WMP300N

注：装好无线网卡以后，务必记得开 PC4 的电源（之前把它关闭了）。PC4 就会连接到无线 AP，如图 2.46 所示。

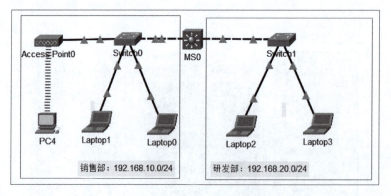

图 2.46　无线连接成功

步骤 3：为 PC4 配置 IP 地址（192.168.10.14/24）和网关（192.168.10.1），如图 2.47 所示。

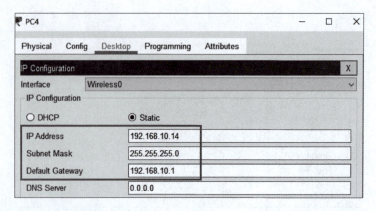

图 2.47　为 PC4 配置 IP 地址（192.168.10.14/24）和网关（192.168.10.1）

步骤 4：在 Switch0 上，将无线 AP 接入的端口（Fa0/23）划入销售部所在的虚拟局域网（VLAN 10），命令如下：

```
Switch0>enable    //进入特权模式
Switch0#configure terminal    //进入全局配置模式
Switch0(config)#interface fastEthernet 0/23    //进入无线 AP 占用的端
口(Fa0/23)
Switch0(config-if)#switchport mode access    //设置端口的访问模式
Switch0(config-if)#switchport access vlan 10    //将端口划入 VLAN 10
Switch0(config-if)#exit
Switch0(config)#
```

查看 Switch0 的 VLAN 数据库信息，以便确认是否划入成功，结果如图 2.48 所示。

```
Switch0#show vlan

VLAN Name                             Status    Ports
---- -------------------------------- --------- -------------------------------
1    default                          active    Fa0/2, Fa0/3, Fa0/5, Fa0/6
                                                Fa0/7, Fa0/8, Fa0/9, Fa0/10
                                                Fa0/11, Fa0/12, Fa0/13, Fa0/14
                                                Fa0/15, Fa0/16, Fa0/17, Fa0/18
                                                Fa0/19, Fa0/20, Fa0/21, Fa0/22
                                                Gig0/1, Gig0/2
10   Sales                            active    Fa0/1, Fa0/4, Fa0/23
1002 fddi-default                     active
1003 token-ring-default               active
1004 fddinet-default                  active
1005 trnet-default                    active
```

图 2.48　Switch0 的 VLAN 数据库信息

可以看到 Fa0/23 已经成功划入 VLAN 10 中。

步骤 5：测试连通性。

从 PC4 ping 研发部的主机 Laptop2，结果如图 2.49 所示。

```
C:\>ping 192.168.20.12

Pinging 192.168.20.12 with 32 bytes of data:

Reply from 192.168.20.12: bytes=32 time=41ms TTL=127
Reply from 192.168.20.12: bytes=32 time=27ms TTL=127
Reply from 192.168.20.12: bytes=32 time=20ms TTL=127
Reply from 192.168.20.12: bytes=32 time=13ms TTL=127

Ping statistics for 192.168.20.12:
    Packets: Sent = 4, Received = 4, Lost = 0 (0% loss),
Approximate round trip times in milli-seconds:
    Minimum = 13ms, Maximum = 41ms, Average = 25ms
```

图 2.49　从 PC4 ping Laptop2 的结果

从测试结果看到的信息判断，PC4 和 Laptop2 之间的通信是正常的。说明 PC4 可以通过无线的方式成功接入局域网，并实现良好的数据通信。

这里只是介绍另一种网络接入的方式——无线。实际上现在市场上在售的笔记本和部分台式机都已经配置无线网卡，不用另外购买；另外，无线 AP 还需要更多的配置，以提供更安全的访问，但由于思科的模拟器在 AP 操作上与实际设备区别比较大，在这里不做过多演示。

课堂练习：在网络中部署无线 AP，让用户可以通过无线访问网络。

素质拓展：无线技术用于数据传输对人类社会无疑是一场革命，它再次解放了人类。但是事物总是矛盾的，无线技术固有的缺陷也影响了数据传输的安全性。要提高信息安全意识，通过无线网络传输数据时，要防止敏感信息泄露。更不能利用无线技术的安全缺陷进行非法的数据窃取，任何组织和个人都要在宪法和法律范围内活动，一切违法行为都应受到法律的追究。

小结与拓展

1. 交换机、路由器、无线 AP 的应用场合

如果通信的发送方和接收方处于同一个逻辑网络（网络地址一样），则用交换机互连；如果通信的发送方和接收方处于不同的逻辑网络（网络地址不一样），则用三层设备（三层交换机或路由器）作为中间转发设备；无线 AP 则是为移动用户提供 WiFi 接入的应用场景。

2. 思科网络设备 3 种常用模式的区别

思科路由器和交换机的命令行有 3 种模式：用户模式、特权模式、全局配置模式（其实还有各种子配置模式，在这里不作详细讨论）。

不同的模式是按照具备的权限和作用来区分的。

普通的用户模式是权限最小的模式。设备刚开始启动进去的时候，就是用户模式。只能查看一些基础信息和调用几个基本的网络测试工具，如 ping、traceroute 等。

在用户模式下运行"enable"命令就可以进入到特权模式。特权模式具有最高的权限，几乎可以对设备进行所有的操作。

在特权模式下运行"configure terminal"命令可以进入全局配置模式。顾名思义，这个模式就是为工程师提供配置的环境，几乎所有的配置都需要在这个模式及其子模式下操作。

每个模式下能做的事情是不一样的，通常情况下我们需要在不同的模式下切换以便完成一项任务。刚开机进去的模式是用户模式，想要配置设备就要在全局配置模式或其子模式下进行，可是不能从用户模式直接切换到全局配置模式，需要先进入特权模式，再进入全局配置模式。网络设备的不同模式可以类比为家的不同房间，每个房间的功能是不一样的，比如说我想回家睡觉，进门先到客厅，再进入卧室去睡觉，而不能直接从窗户爬进卧室去睡觉。

3. 思科网络设备实用命令及技巧

在对设备进行配置的过程中，可以用"?"或"Tab"键来辅助完成命令的查询或输入。

可以使用问号（?）来查询在该状态下可以支持的命令。比如要查询特权模式下可以执行哪些命令，可以在当前状态下输入"?"，结果如图 2.50 所示。

每条可执行的命令后面还注明了该命令的功能。可以通过查询厂家关于产品的技术文档了解命令的使用。

```
MS0>?
Exec commands:
  <1-99>       Session number to resume
  connect      Open a terminal connection
  disable      Turn off privileged commands
  disconnect   Disconnect an existing network connection
  enable       Turn on privileged commands
  exit         Exit from the EXEC
  logout       Exit from the EXEC
  ping         Send echo messages
  resume       Resume an active network connection
  show         Show running system information
  ssh          Open a secure shell client connection
  telnet       Open a telnet connection
  terminal     Set terminal line parameters
  traceroute   Trace route to destination
MS0>
```

图 2.50　用"?"查询可执行的命令

也可以查询以某些可以确定的字符串开头的命令。比如查询以"dis"开头的命令，操作如图 2.51 所示。

根据查到的信息，在该模式下，以"dis"开头的命令有两个，disable 和 disconnect。

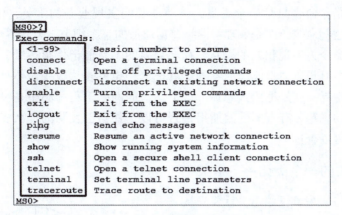

```
MS0#
MS0#dis?
disable    disconnect
MS0#dis
```

图 2.51　查询以"dis"开头的命令

Tab 键可以帮助工程师在输入的命令字符串唯一的条件下补全命令。字符串唯一是什么意思呢？如图 2.51 所示的案例，以 dis 开头的命令有两个，就不是唯一的，此时按 Tab 无法补全命令，因为设备不知道你要补的是 disable 还是 disconnect。但是如果输入"disa"然后再按 Tab 键，系统就自动帮我们把命令补齐，显示完整的"disable"；如果输入"disc"然后再按 Tab 键，系统就自动帮我们把命令补齐，显示完整的"disconnect"，如图 2.52 所示。

```
MS0>disc
MS0>disconnect
```

图 2.52　使用 Tab 键补全命令

其实在输入的命令字符串唯一的条件下，哪怕命令不全，也是可以执行的。因为系统发现以这个字符串开头的命令只有一个，具有唯一性。熟练以后，大多数情况下我们都是这样去操作的，没有必要每条命令都完整输入，以提高效率，命令如下：

```
MS0>
MS0>en        //enable 的简写
MS0#conf t    //configure terminal 的简写
MS0(config)#int vlan 10      //interface 的简写
MS0(config-if)#ip add 192.168.10.1 255.255.255.0    //ip address 的简写
MS0(config-if)#no sh       //no shutdown 的简写
MS0(config-if)#ex       //exit 的简写
MS0(config)#ex        //exit 的简写
MS0#
```

需要注意的是：命令只是一个工具，是辅助我们实现业务的工具，要学会灵活运用。不要去死记硬背这些命令，而是要理解网络数据通信的整个细节，要培养解决问题的思路，然后才想着要用命令去实现我们的思路。社会需要的不是能记住很多纸、笔和颜料的工具人，而是需要一个有想象力，能借助各种工具把那种想象表达成一份作品的艺术家。张三丰说武功最高的境界，就是把招式忘掉。一个伟大的网络工程师心中是没有命令的，只有问题和解决问题的思路。有人问爱因斯坦是否可以记住自己发明的这么多公式，爱因斯坦说：记不住，但可以查！

4. 关于网关的一个类比——边关

数据要从网络 A 发到网络 B，就要先从网络 A 的网关进入，再由网络 A 的网关转发给网络 B 的网关（如果网络 A 和网络 B 是邻接网络），最后从网络 B 的网关出来，转发给网络 B 内的接收方。网关示意图如图 2.53 所示。

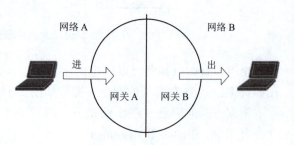

图 2.53　网关示意图

互联网中的网关，其实可以比较通俗地类比成国家之间的边关。如图 2.54 所示，楚汉以河为界，两国之间有两个口岸：AC 和 BD，A，B 是楚的边关，C，D 是汉的边关。楚人入汉可以走 AC 也可以走 BD，但不管怎样，都要从自己国家的边关出境，过了河到对岸，从汉的边关入汉。任何不按照规则越过国境的方式叫偷渡，是不被允许的。

从一个网络到另外一个网络，可能有很多通道，就比如由楚入汉，有多个口岸，从任何一个口岸出去，都可以到达目的地。

类似的，从一个网络到另外一个网络可以有多个网关，我们在用户计算机的网卡上配置的网关的 IP 地址是谁，那么发送方就会把数据发送到这个 IP 地址对应的节点（网关），由它负责转发到另外一个网络。

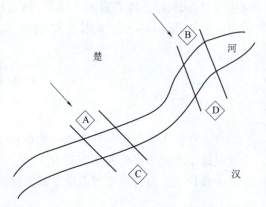

图 2.54　楚汉边境

用楚汉边界来做类比，楚人入汉走 AC 还是 BD，取决于他大脑中记住了哪个口岸。如果他只记得 AC 口岸，那么就从 AC 口岸入汉，如果他只记得 BD 口岸，那么就从 BD 口岸入汉；但是如果这个楚人知道有两个口岸都可以入汉，则轮着走，即这次从 AC 入汉，下次从 BD 入汉，轮完了再反复，这叫负载均衡。

当然，如果楚人哪个口岸都不知道，那它就无法走出楚国。这就像我们没有给用户计算机配置网关 IP 地址一样，用户的计算机就不知道网关是谁，于是找不到任何节点来帮助用户转发数据，数据当然就无法到达接收方。

5. 交换机转发数据的机制——MAC 地址表

交换机根据缓存中的媒体访问控制（Media Access Control，MAC）地址表来决定数据该从交换机的哪个端口转发出去。我们可以通过 show mac-address-table 命令来查看交换机内部的 MAC 地址表，如图 2.55 所示。

```
Switch1#show mac-address-table
        Mac Address Table
-------------------------------------------
Vlan    Mac Address       Type        Ports
----    -----------       --------    -----
 1      0001.c774.3002    DYNAMIC     Fa0/24
20      0001.42d1.e839    DYNAMIC     Fa0/12
20      0060.5c74.2c99    DYNAMIC     Fa0/13
```

图 2.55　Switch1 的 MAC 地址表信息

我们以 Laptop2 发送数据给 Laptop3 为例进行解析。Laptop2 把数据发到交换机 Switch1 后，Switch1 就像一个快递员一样，去读取数据帧上接收方 MAC 地址（目的 MAC 地址），如图 2.56 所示。

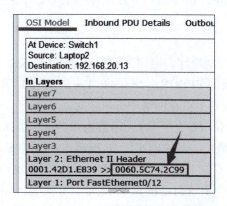

图 2.56　Laptop2 发送 Laptop3 的数据帧

然后去查 MAC 地址表，如图 2.57 所示。

```
Switch1#show mac-address-table
        Mac Address Table
-------------------------------------------
Vlan    Mac Address       Type        Ports
----    -----------       --------    -----
 1      0001.c774.3002    DYNAMIC     Fa0/24
20      0001.42d1.e839    DYNAMIC     Fa0/12
20      0060.5c74.2c99    DYNAMIC     Fa0/13
```

图 2.57　Switch1 的 MAC 地址表信息

发现接收方接入的端口号是 Fa0/13，同时接收方跟自己同属 VLAN 20，于是将发送给 Laptop3 的数据从 Fa0/13 发出去。

需要注意的是，交换机将数据发出去以后，工作就完成了。至于 Laptop3 能不能收到数据，交换机是不管的；甚至交换机根本不会去查证 Laptop3 这台主机是否存在，只要在 MAC 地址表里查到记录，就根据记录中的端口把数据转发出去。

另外，如果交换机在 MAC 地址表中查询不到接收方的 MAC 地址记录，该如何处理？答案是：交换机将通过泛洪的方式将这个数据帧发出去（交换机的每个端口发一份，希望有一个端口是对的）。这种方法在一定程度上会造成交换网络中数据泛滥，影响网络通信效率，但至少也算是一种努力尝试转发的方式，以减少丢包。

那么这个 MAC 地址表里的记录是怎么来的呢？要怎样才可以在表中写入一条记录呢？最简单的方法是，交换机会通过"守株待兔"的方式去自己学习。MAC 地址表实际上记录的是某台主机占用哪个端口，这个端口被划入哪个 VLAN。MAC 地址表保存在缓存内，意味着交换机刚开机的时候这个 MAC 地址表是空的，是没有记录的。但是交换机会不停地去监听每个端口，当有数据从某个端口进来的时候，交换机就捕获数据，记录下数据帧上发送方的 MAC 地址，并将该 MAC 地址和捕获的端口号对应在一起，记录在 MAC 地址表中，表示某台主机占用的是这个端口。

MAC 地址表因为是临时生成的，因此保存在缓存中，在交换机断电后就会丢失。其实我们也可以通过 clear mac-address-table 命令在不断电的状态下清空交换机的 MAC 地址表。

6. MAC 地址

MAC 直译为介质访问控制地址，也称为物理地址（Physical Address），是由网络设备制造商生产时烧录在硬件内部的 EPROM（Erasable Programmable Read-Only Memory）芯片上的一个参数，这个参数代表着主机在网络中的唯一标志。

MAC 地址的长度为 48 位（6 个字节），通常表示为 12 个 16 进制数，如：00-16-EA-AE-3C-40 就是一个 MAC 地址。其中前 6 位 16 进制数 00-16-EA 代表网络硬件制造商的编号，它由 IEEE 分配，而后 6 位 16 进制数 AE-3C-40 代表该制造商所制造的某个网络产品（如网卡）的系列号。只要不更改自己的 MAC 地址，MAC 地址在世界就是唯一的。形象地说，MAC 地址就如同身份证上的身份证号码，具有唯一性。

如果不用非常规手段去修改它，那么 MAC 地址是固定的，不会随着设备处于不同的网络而发生变化。与 MAC 地址不一样的是，IP 地址则会随着设备迁移到不同的网络，必须要获得该网络的主机地址才能参与网络通信。从这个角度去比较，MAC 地址可以类比我们的身份证号，将伴随着一生；而 IP 地址就类似人的住址，处于不同的城市，街道名称和门牌号是不一样的。

7. 控制 VLAN 流量通过 Trunk 链路

对于思科的交换机，默认情况下允许所有 VLAN 的流量通过 Trunk 链路，如果要对通过 Trunk 链路的 VLAN 流量进行管控，则用以下命令可以实现只允许某些特定的 VLAN 流量通过 Trunk 链路：

```
    Switch1(config)#int fa0/24         //进入 FastEthernet 0/24 端口
    Switch1(config-if)#switchport mode trunk     //将端口的工作模式设置为 Trunk(干道模式)
    Switch1(config-if)#switchport trunk allowed vlan 10,20     //只允许 VLAN 10 和 VLAN 20 的流量通过
```

通过 show interfaces trunk 命令可以查看允许通过某个 Trunk 端口的 VLAN，结果如图 2.58 所示。

```
Switch1#show interfaces trunk
Port      Mode           Encapsulation    Status        Native vlan
Fa0/24    on             802.1q           trunking      1

Port      Vlans allowed on trunk
Fa0/24    10,20

Port      Vlans allowed and active in management domain
Fa0/24    10,20
```

图 2.58　查看允许通过某个 Trunk 端口的 VLAN

发现交换机 Switch1 的 FastEthernet 0/24 端口，只允许 VLAN 10 和 VLAN 20 的流量通过。但是如果想要拒绝某些特定的 VLAN 流量通过，则可以用以下命令实现：

Switch1(config)#int fa 0/24   //进入 FastEthernet 0/24 端口
Switch1(config-if)#switchport mode trunk   //将端口的工作模式设置为 Trunk(干道模式)
Switch1(config-if)#switchport trunk allowed vlan except 10   //除 VLAN 10 的其他流量都被允许通过

再次通过 show interfaces trunk 命令查看 Switch1 的 FastEthernet 0/24 端口的状态，结果如图 2.59 所示。

```
Switch1#sh interfaces trunk
Port      Mode           Encapsulation    Status        Native vlan
Fa0/24    on             802.1q           trunking      1

Port      Vlans allowed on trunk
Fa0/24    1-9,11-1005

Port      Vlans allowed and active in management domain
Fa0/24    1,20
```

图 2.59　查看 Switch1 的 Fa0/24 端口的状态

从显示的结果可以看到，VLAN 1~9、VLAN 11~1005 的流量都报备允许通过 Trunk 端口（Fa0/24）。

8. 三层设备（路由器或三层交换机）转发数据的机制——IP 路由表

三层设备根据缓存中的 IP 路由表来决定数据该从哪个端口转发出去。我们可以通过 show ip route 命令来查看路由表，如图 2.60 所示。

这个路由表中有两条记录，记录了什么信息呢？这里解读第一个路由条目。首先第一个字段"C"表示这是一个直连网络，也就是说这两个网络是直接连在这台设备上的（其他的代码表示的信息在上方有说明）；接着第二个字段是网络号（192.168.10.0/24），还说明了这个网络是直接连接的（is directly connected）；最后一个字段表示发往 192.168.10.0/24 的网络数据要从 VLAN10 端口转发出去。

如研发部的 Laptop3（192.168.20.13/24）有数据要发到销售部 Laptop1（192.168.10.11/24），则 Laptop3 先把数据包发到网关（VLAN20：192.168.20.1/24）；MS0 从 VLAN20 端口接收

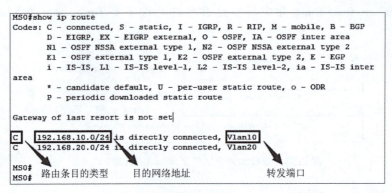

图 2.60 MS0 的路由表

到数据以后,也会像个快递员一样,去读取这个数据包上的目的 IP 地址(192.168.10.11);然后 MS0 去查路由表里的条目,会匹配到网络地址 192.168.10.0/24 的条目(如果路由表中有多条 192.168.10.0 网络的条目,则匹配到子网掩码最长的那个路由条目,这就是最长匹配原则),根据匹配到的路由条目的要求,这个数据将通过 VLAN10 端口发出。

跟交换机一样,路由器将数据包转发出去以后,就完成任务了。它不会去管接收方是否会接收到,甚至不知道接收方是否真的存在。只要路由条目里写的该从哪个端口转发出去,它就严格按照路由表中路由条目的要求去转发。

与交换机转发数据帧不一样的是,路由器去查询路由表,如果查不到目的网络的路由条目,则会把这个数据包丢弃。

你可以想象,一个快递员看到一个快递上的目的地址,努力扫遍整个脑海都查不到关于这个地址的任何信息,然后就把这个包裹丢弃!这在我们看来是一件很滑稽的事情,可是路由器就是这样转发数据的,而且符合规则。

路由表的路由条目是怎么来的呢?当然也有很多种方式,但大概可以分 3 类:第 1 类叫静态路由,是网络管理员手动配置进去的,用 "S" 标记,表示 "static" 的意思;第 2 类叫动态路由,就是三层设备通过路由协议(RIP、OSPF、BGP 等)跟邻居学习到的,或者是通过邻居发来的信息后自己算出来的;第 3 类就是本案例的路由条目,就是自己直接连接的,用 "C" 标记,路由信息就在自己的缓存中,不需要别的设备发给它。

9. A 类,B 类,C 类 IP 地址默认的子网掩码

我们在给计算机配置 IP 地址的时候,系统会自动给我们分配一个默认的子网掩码。它根据表 2.10 确定这个默认的子网掩码。

表 2.10 各类 IP 地址默认子网掩码

网络类型	默认子网掩码	掩码长度
A	255.0.0.0	/8
B	255.255.0.0	/16
C	255.255.255.0	/24

10. 无线局域网（WLAN）基础

无线局域网是不使用任何导线或传输电缆连接的局域网，而使用无线电波或电场与磁场作为数据传送的介质，传送距离一般只有几十米。但是无线局域网的主干网络通常使用有线电缆，无线局域网用户通过一个或多个无线 AP 接入无线局域网。

无线局域网第一个版本 IEEE 802.11 发表于 1997 年，其中定义了介质访问接入控制层和物理层。物理层定义了工作在 2.4 GHz 的 ISM 频段上的两种无线调频方式和一种红外传输的方式，总数据传输速率设计为 2 Mbit/s。经过几十年的发展，无线局域网技术已经广泛地应用在商务区、大学、机场及其他需要无在线网的公共区域。

通常情况下，很多人认为 WLAN 就是 WiFi。需要说明的是，它们不是同一个概念。WLAN 的标准叫 IEEE 802.11，WiFi 只是 IEEE 802.11 标准的一种实现，只是对于普通用户来说，WiFi 使用得最普遍。基于 IEEE802.11 标准的产品除了 WiFi，还有无线千兆（Wireless Gigabit，WiGig）联盟。WiGig 联盟于 2013 年 1 月 4 日并入 WiFi 联盟。

目前 WiFi 已发展到第六代，如表 2.11 所示。

表 2.11　WiFi 发展世代表

世代	年份	依据的标准	工作频段	最高速率
第一代	1997	IEEE 802.11 原始标准	2.4 GHz	2 Mbit/s
第二代	1999	IEEE 802.11b	2.4 GHz	11 Mbit/s
第三代	1999	IEEE 802.11a	5 GHz	54 Mbit/s
第四代	2009	IEEE 802.11n	2.4 GHz 和 5 GHz	600 Mbit/s
第五代	2013	IEEE 802.11ac	5 GHz	6.9 Gbit/s
第六代	2019	IEEE 802.11ax	2.4 GHz 和 5 GHz	9.6 Gbit/s

11. 保存设备的配置

必须要记得保存设备的配置信息，以免丢失。

配置设备的时候，设备的信息保存在一个名为 running-config 的配置文件中，而这个文件存放在网络设备的内存中，是临时的，断电后会丢失。设备中还有一个名为 startup-config 的配置文件，存放在设备的 flash 芯片中，属于永久保存的文件。

我们在完成配置以后，要在特权模式下用 copy running-config startup-config 将配置信息从临时的 running-config 复制到 startup-config，或者用 write 命令将配置信息保存起来，防止断电后配置信息丢失。

12. 在非特权模式下，强制执行特权模式下的命令

在配置模式，我们无法直接运行诸如 show 或 ping 等特权模式下的命令，但是可以用 do 强制执行。

每个模式下都有相应的命令，我们通常需要一边在配置模式配置设备，一边要用 show 检查配置情况以及用 ping 命令测试通信效果。可是如果从配置模式退出到特权模式去 show 或者 ping，会降低工作效率，此时可以直接在配置模式下，用 do 命令强制执行特权模式下

的命令。

比如在配置模式下进行保存设备配置的操作时,可通过 do write 命令完成。

思考与训练

(1) 有一台主机的 IP 地址是 172.16.1.1/28,请问该主机所在网络的网络地址、广播地址、可用的主机地址分别是什么?

(2) IP 地址 172.16.1.16/28 是否可以分配给主机使用?为什么?地址 172.16.1.15/28 呢?

(3) 主机 211.16.1.1/30 和主机 211.16.1.5/30 是否在同一个逻辑网络内?如果它们要相互通信,必须经过什么设备帮忙转发数据?

(4) 交换机转发数据的依据是什么?如何转发?请用自己的语言进行描述。

(5) 路由器转发数据的依据是什么?如何转发?请用自己的语言进行描述。

(6) 无线局域网就是 WiFi,这种说法正确吗?为什么?

(7) 参照如图 2.61 所示的校园网简单拓扑图,用 Cisco Packet Tracer 设计一个网络系统,适当配置,让全网用户能相互通信并且能访问到内网的服务器。图中××表示自己学号最后两位,根据给定的 IP 地址段规划主机和网关的 IP 地址;端口号、VLAN 号和名称自行决定。

关键提示:

①需要创建 3 个 VLAN,将教学楼、宿舍楼、服务器群划分在 3 个不同的逻辑网络中;

②根据给定的网络地址计算每个逻辑网络可用的主机地址范围,并根据这些地址为每个网络内的主机和网关分配 IP 地址;

③注意三层交换机 MS0 的路由功能要手动开启;

④核心技能是培养解决问题的思路。(计算主机地址→分配 IP 地址→创建 VLAN→将端口划入相应 VLAN→配置 Trunk 链路→核心交换机配置 SVI 端口地址作为网关→启用路由功能)

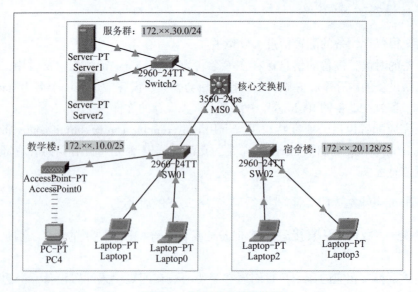

图 2.61 校园网简单拓扑图

Project 2

Expanding the Network

A practical computer network certainly involves more than two hosts. This project expands the number of computers in a network and addresses various resulting issues. Upon completion, learners will understand how a multitude of computers can interconnect, how to determine if these computers belong to the same or different networks, and how to enable communication between computers on different networks with WLAN technology.

Learning Objectives

(1) Understand the role of switches and routers in a network;
(2) Understand the significance and function of subnet masks;
(3) Grasp the meaning and role of a gateway;
(4) Learn the basic working principles of switches and routers;
(5) Understand the definition and basic knowledge of Virtual Local Area Networks (VLAN);
(6) Gain knowledge about wireless networks.

Skill Objectives

(1) Master command-line configuration methods for switches and routers;
(2) Calculate the range of logical network IP addresses (network, broadcast, and available host addresses);
(3) Learn methods to communicate between different logical networks;
(4) Implement WLAN to enable network access for mobile users.

Relevant Knowledge

In Task 1 of Project 1, the network topology involves only two hosts and is somewhat limited in scope. It simply connects two computers with a crossover twisted pair cable, allowing only these two hosts to access each other in this basic network setup. However, in an office environment, where several or even dozens of people work together, a more complex solution is needed to connect many computers. People might intuitively think of a fully interconnected topology for this purpose, as illustrated in Figure 2.1.

Such a topology is quite intricate and expensive because it requires to equip each computer with numerous network cards to provide enough ports for connecting to other computers. Additionally, this method is not feasible for interconnecting the hundreds of millions of computers globally.

The star or extended star topology, as described in Project 1, offers a solution. It involves a central network device with multiple ports that connects surrounding computers. This central device, responsible for data

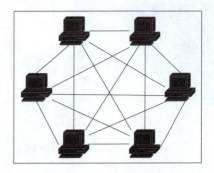

Figure 2.1　Fully interconnected topology

forwarding among terminal devices, is known as an intermediary device in the network. Previously, LANs used devices like hubs, bridges, and switches for data forwarding. Hubs, which allow only one host to send data at a time, were slow and inefficient. Bridges improved upon this by isolating collision domains and using MAC address tables for forwarding, but their application was limited initially due to only two ports.

Following the development of hubs and bridges, Kalpana Company (acquired by Cisco in 1994) released the first Ethernet switch. Essentially, it integrates multiple bridges into one device, functioning as a bridge with multiple ports. This integration method is commonly referred to as "bridging," and even today, Ethernet switches are sometimes called "bridges" in English documents.

Currently, Ethernet switches are the most commonly used intermediary devices in local area networks (LANs). The next section will demonstrate how to use Cisco Packet Tracer to interconnect multiple computers with an Ethernet switch.

Task 1：Building a Small Local Area Network—Switch

Task Objective：Interconnect multiple computers with an Ethernet switch.

Step 1：Drag out a 2,960 switch (Switch) and four computers (Laptop) in the Cisco Packet Tracer workspace, as shown in Figure 2.2.

Building a Small Local Area Network—Switch

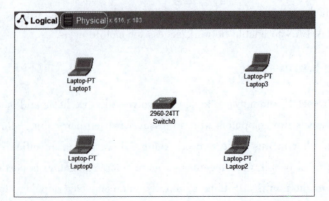

Figure 2.2　Dragging out the switch and computer

Step 2: Connect the laptops to the switch by using straight-through twisted pair cables (Copper Straight-Through) as shown in Figure 2.3.

Note: You can connect to any FastEthernet port on the switch.

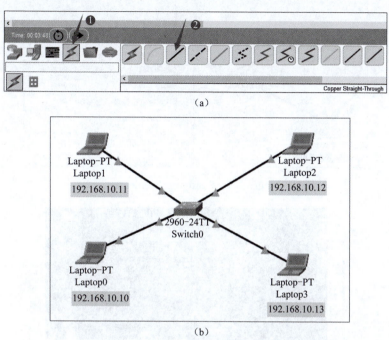

Figure 2.3 Connecting a laptop to a Switch

(a) Choose direct twisted pair cable; (b) Connect to a Switch

Step 3: Configure the IP addresses for the four laptops as per Table 2.1, using default subnet mask values.

Table 2.1 IP address allocation table

Host name	IP address	Notes
Laptop0	192.168.10.10	default subnet mask
Laptop1	192.168.10.11	default subnet mask
Laptop2	192.168.10.12	default subnet mask
Laptop3	192.168.10.13	default subnet mask

Set the IP address for Laptop0 as shown in Figure 2.4.

Use the same method to configure the IP addresses for the other three hosts.

Step 4: Testing

Ping from Laptop0 to Laptop1; the result should be what's shown in Figure 2.5.

Pinging between other hosts should also be successful.

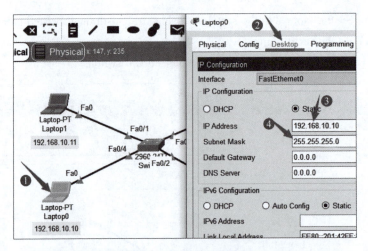

Figure 2.4　Configure IP address

Figure 2.5　Ping test results

Exercise: Following this task, use a switch to connect four computers, configure their IP addresses with default subnet masks, and enable communication between different computers.

Introduction of ideological and political education: Deploying a small Local Area Network (LAN) fosters improved conditions for team collaboration. Team spirit has become one of the most crucial soft skills in career development, making it particularly important to cultivate a correct collective concept in learning and work. Teamwork is built on the foundation of the team, utilizing team spirit and mutual support to achieve maximum work efficiency. Team members need not only individual capabilities but also the ability to contribute to various roles and coordinate with others effectively.

Task 2: Identifying the Network of a Computer—Subnet Mask

Identifying the Network of a Computer—Subnet Mask

Task Objective: Understand subnet masks and master how to calcalate a host's network address, broadcast address, and the range of available host addresses.

When configuring IP addresses, clicking on the subnet mask text box automatically populates the subnet mask.

Let's do a test on Windows: delete the subnet mask parameter of the network card, and then click the "Certain" button. The resulting prompt, as shown in Figure 2.6, indicates that a subnet mask is required to set up an IP address, underscoring its necessity.

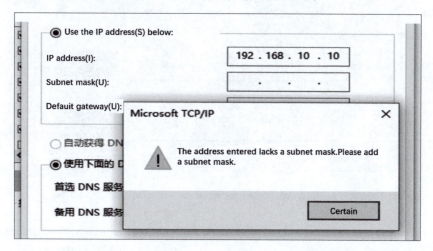

Figure 2.6　Prompt message after deleting subnet mask

1. Understanding Subnet Masks

Like an IP address, a subnet mask is represented in dotted decimal notation. It's a bitmask that distinguishes which bits of an IP address represent the network address and which represent the host address. Subnet masks identify the network/host parts of an IPv4 address and are essentially a sequence of 1s followed by a sequence of 0s.

There are two key characteristics of subnet masks:

(1) They are composed of 32 binary bits, like IP addresses, but the 1s and 0s must be contiguous without crossovers, as shown in Figure 2.7.

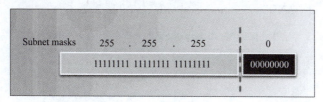

Figure 2.7　Decimal and binary representation methods for subnet masks

Subnet masks have a more convenient expression method used commonly, which is to denote the mask by the number of '1's in its binary form. For instance, the subnet mask 255.255.255.0, shown in binary in Figure 2.7, has 24 '1's. Hence, this subnet mask can be represented as "/24". This notation is widely used because of its convenience.

(2) Subnet masks are used in conjunction with IP addresses to distinguish which parts of the 32-bit IP address are network bits and which are host bits.

As shown in Figure 2.8, the '1's in a subnet mask correspond to the network portion of the

IP address, and the '0's correspond to the host portion.

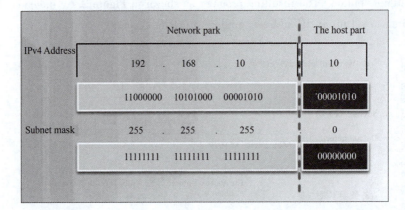

Figure 2.8 Correspondence between subnet mask and IP address

It's important to note that the subnet mask itself does not contain any network information from the IP address; it only informs the host of which parts of its 32-bit binary IP address are network bits and which are host bits. Without a subnet mask, a host cannot determine its network and host bits. The network bits are essential for identifying the network; without clarity, the host's identity becomes ambiguous, leading to difficulties in data transmission. This is analogous to a person who cannot send or receive parcels if he doesn't know his own location.

2. Calculating the Network Address, Broadcast Address, and Range of Available Host Addresses for a Host

Knowing the IP address and its subnet mask allows us to determine the network to which the host belongs. The method is simple: convert the IP address and subnet mask into binary and perform a logical "AND" operation.

The binary logical "AND" operation is shown in Table 2.2.

Table 2.2 Binary logical and operation table

Logic variable	Logical operator	Logic variable	Result
1	AND	1	1
0	AND	1	0
0	AND	0	0
1	AND	0	0

Note: The logical AND operation can be analogized to multiplication in arithmetic, as they yield the same result.

We'll use an example to understand how to calculate the network address, broadcast address, and the range of available host addresses for a host.

Example: A host has the IP address 192.168.10.10 with a subnet mask of 255.255.255.0.

Calculate the network's network address, broadcast address, and the range of available host addresses.

Step 1: Convert the decimal IP address and subnet mask to binary, as shown in Table 2.3.

Table 2.3 Decimal and binary correspondence of IP address and subnet mask

IP address	192	168	10	10
Subnet Mask	255	255	255	0
Binary IP address	11000000	10101000	00001010	00001010
Binary subnet mask	11111111	11111111	11111111	00000000

Step 2: Logical "AND" operation between the IP address and the subnet mask reveals the network address of the host's network, as shown in Table 2.4.

Table 2.4 Logical operation table for binary IP addresses and binary subnet masks

Binary IP address	11000000	10101000	00001010	00001010
Logical operation		AND		
Binary subnet mask	11111111	11111111	11111111	00000000
Result	11000000	10101000	00001010	00000000

The network address for a host with the IP 192.168.10.10/24 is represented in binary as 11000000.10101000.00001010.00000000/24, with all host bits being '0'. This translates to 192.168.10.0/24 in decimal, indicating the smallest address in the network, which is unique on the Internet.

Note: Always include the subnet mask when expressing an IP address.

Step 3: Replacing all host bits of the network address with '1' gives the broadcast address. In decimal, this is 192.168.10.255/24, the largest address in that network, as shown in Figure 2.9.

	Network bit			Host position
Binary IP address	11000000	10101000	00001010	00001010
Logical operation		AND		
Binatry subnet mask	11111111	11111111	11111111	00000000
Network address	11000000	10101000	00001010	00000000
Broadcast address	11000000	10101000	00001010	11111111

Figure 2.9 Broadcast address calculation process

Step 4: Determine the range of IP addresses available for allocation to hosts. Following these steps, the network and broadcast addresses for a host with the IP 192.168.10.10/24 are shown in Table 2.5.

Table 2.5 Network address and broadcast address table

Address type	Value	Notes
Network address	192.168.10.0/24	The minimum IP address in the network where the host is located
Broadcast address	192.168.10.255/24	The maximum IP address in the network where the host is located

By knowing the minimum and maximum IP addresses, a range can be obtained, as shown in Figure 2.10.

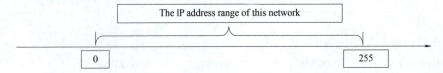

Figure 2.10　IP address coordinate range diagram

In the network, the available host addresses should be within the range of 0-255. However, 192.168.10.0/24 is used as the network address and 192.168.10.255/24 as the broadcast address. Therefore, the range 192.168.10.1 to 192.168.10.254 is available for the host allocation. It's important to note that 192.168.10.0/24 and 192.168.10.255/24 are reserved for the network and broadcast addresses, respectively, and cannot be assigned to hosts.

A subnet mask is used to determine the network and host parts of an IP address, identifying which logical network it belongs to. Computers on the Internet are in different logical networks, but they must follow network communication rules for interconnectivity.

Exercise: Use the "ipconfig" command to view your computer's IP address and subnet mask, and calculate the network address, broadcast address, and the host range available for your network. Compare with a classmate to see if your computers are on the same network.

Introduction of ideological and political education: The essence of computer network data communication lies in hosts adhering to common rules, emphasizing the importance of order and regulations. Similarly, university students should understand legal norms and school regulations thoroughly, establish a strong sense of rule of law, take pride in abiding by laws and disciplines, and consciously adhere to relevant disciplines and laws.

Task 3: Connecting Different Networks with a Router—Gateway

Connecting Different Networks with a Router—Gateway

Task Objective: Understand the role of routers and their basic configurations to enable communication between different networks.

1. Design A Topology and Test Communication Between Hosts on Different Networks

In Project 1, Task 1, the four hosts' addresses range from 192.168.10.10 to 192.168.10.13/24, all sharing the network address 192.168.10.0/24, indicating they are in the same network.

What if the IP addresses are assigned as shown in Table 2.6? Would communication between the hosts still be normal?

Table 2. 6 IP address allocation table and its network address

Host name	IP address	Network address	Notes
Laptop0	192. 168. 10. 10/24	192. 168. 10. 0/24	Same network
Laptop1	192. 168. 10. 11/24	192. 168. 10. 0/24	
Laptop2	192. 168. 20. 12/24	192. 168. 20. 0/24	Same network
Laptop3	192. 168. 20. 13/24	192. 168. 20. 0/24	

Test the communication between hosts within the same network. Laptop0 ping Laptop1, and the results are shown in Figure 2. 11.

```
C:\>ping 192.168.10.11

Pinging 192.168.10.11 with 32 bytes of data:

Reply from 192.168.10.11: bytes=32 time=1ms TTL=128
Reply from 192.168.10.11: bytes=32 time<1ms TTL=128
Reply from 192.168.10.11: bytes=32 time=1ms TTL=128
Reply from 192.168.10.11: bytes=32 time<1ms TTL=128

Ping statistics for 192.168.10.11:
    Packets: Sent = 4, Received = 4, Lost = 0 (0% loss),
Approximate round trip times in milli-seconds:
    Minimum = 0ms, Maximum = 1ms, Average = 0ms

C:\>
```

Figure 2. 11 Results from Laptop0 ping Laptop1

Laptop2 sends ICMP data to Laptop3 (IP address: 192. 168. 20. 13) using the ping command, with results in Figure 2. 12.

```
C:\>ping 192.168.20.13

Pinging 192.168.20.13 with 32 bytes of data:

Reply from 192.168.20.13: bytes=32 time=1ms TTL=128
Reply from 192.168.20.13: bytes=32 time<1ms TTL=128
Reply from 192.168.20.13: bytes=32 time<1ms TTL=128
Reply from 192.168.20.13: bytes=32 time<1ms TTL=128

Ping statistics for 192.168.20.13:
    Packets: Sent = 4, Received = 4, Lost = 0 (0% loss),
Approximate round trip times in milli-seconds:
    Minimum = 0ms, Maximum = 1ms, Average = 0ms

C:\>
```

Figure 2. 12 Results of ping Laptop3 from Laptop2

Hosts within the same network communicate without any issues. However, when testing communication between devices from different networks (Laptop1 pinging Laptop3), the result is shown in Figure 2. 13.

```
C:\>ping 192.168.20.13

Pinging 192.168.20.13 with 32 bytes of data:

Request timed out.
Request timed out.
Request timed out.
Request timed out.

Ping statistics for 192.168.20.13:
    Packets: Sent = 4, Received = 0, Lost = 4 (100% loss),
```

Figure 2. 13 Results of ping Laptop3 from Laptop1

Next, Laptop0 sends a set of ICMP data to Laptop2 (IP address: 192.168.20.12) using the ping command, with results shown in Figure 2.14.

```
C:\>ping 192.168.20.12

Pinging 192.168.20.12 with 32 bytes of data:

Request timed out.
Request timed out.
Request timed out.
Request timed out.

Ping statistics for 192.168.20.12:
    Packets: Sent = 4, Received = 0, Lost = 4 (100% loss),
```

Figure 2.14 Results of ping Laptop2 from Laptop0

All results indicate "Request timed out," suggesting that the data sent does not reach the recipient, nor is an acknowledgment packet received from Laptop2. This reveals a pattern: communication within the same network is normal, but hosts from different networks cannot communicate. The conclusion is that switches can only forward data within the same network, but not between different ones. For inter-network communication, layer 3 devices like routers or layer 3 switches are needed.

2. Connecting Different Networks with a Router

Next, we'll connect these two networks using the router's Fast Ethernet (Fa) ports, with the modified setup shown in Figure 2.15.

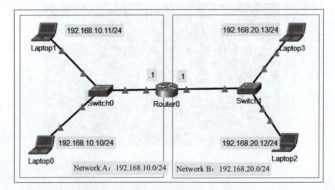

Figure 2.15 Topology after project renovation

In the diagram, Router0's Fa0/0 port serves as the gateway for Network A, handling data traffic forwarding to and from other networks; similarly, the Fa0/1 port is the gateway for Network B.

3. Simple Configuration of Routers

Routers are akin to computers in that each network port must be assigned with an IP address to participate in communication. For the router to function correctly and forward data between different networks, it's essential to configure its ports that connect these networks, typically known as gateways. The IP addresses for these router ports are designated as shown in Table 2.7.

Table 2.7 Router Router0 Port IP Address Allocation Table

Port name	IP address	Notes
Fa0/0	192.168.10.1/24	Must be on the same network as Laptop0 and 1
Fa0/1	192.168.20.1/24	Must be on the same network as Laptop2 and 3

Note that the gateway's IP address must be within the range of host addresses available for its network.

Next, we will begin configuring the router.

Step 1: Use a console cable to connect Laptop4's RS232 port to the router's console port, as depicted in Figure 2.16.

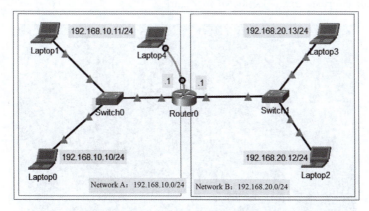

Figure 2.16 Management computer connection device console port

Step 2: Log into the router's console interface using terminal emulation software, such as HyperTerminal, from the administrative computer, as shown in Figure 2.17.

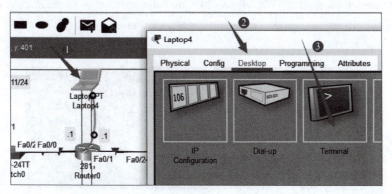

Figure 2.17 Location of hyper terminal software

Keep the connection parameters unchanged and click "OK" button, as shown in Figure 2.18.

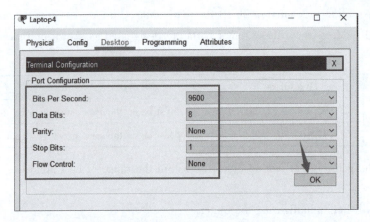

Figure 2.18 Hyper terminal login console parameters

The Terminal interface opens as shown in Figure 2.19.

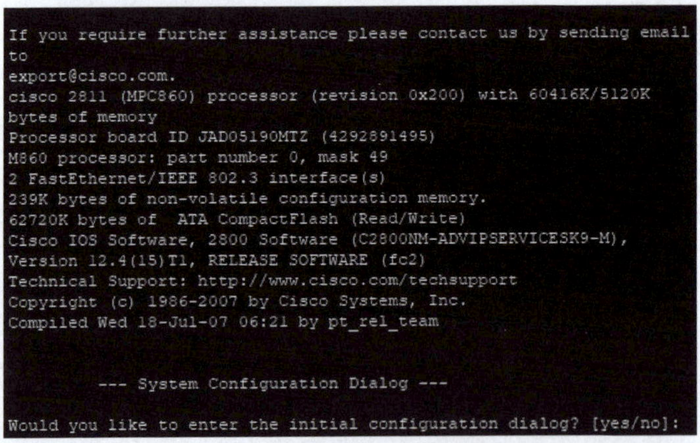

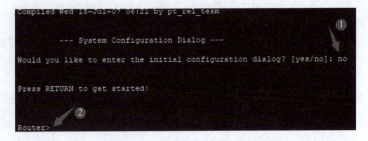

Figure 2.19 Initial interface for logging into the router for the first time

Where the router prompts, "Would you like to enter the initial configuration dialog?" with options [yes/no]. Here, we input "no" and press enter to confirm, moving into user mode as depicted in Figure 2.20.

Figure 2.20 Entering User Mode

Step 3: Configure the router's Fa0/0 port with the IP address 192.168.10.1/24 to serve as the gateway for Network A, providing cross-network data forwarding services for hosts in Network A.

```
Router>enable        //Enter privileged mode
Router#configure terminal     //Enter the global configuration mode
Router(config)#interface fa0/0     //Enter port Fa0/0
Router(config-if)#ip address 192.168.10.1 255.255.255.0     //The
IP address of the port is specified
Router(config-if)#no shutdown     //Active port
Router(config-if)#exit     //Exit port
Router(config)#     //Back to global configuration mode
```

Step 4: Configure the router's Fa0/1 port with the IP address 192.168.20.1/24 to serve as the gateway for Network B, providing cross-network data forwarding services for hosts in Network B.

```
Router(config)#interface fa0/1     //Enter port Fa0/1
Router(config-if)#ip address 192.168.20.1 255.255.255.0     //The
IP address of the port is specified
Router(config-if)#no shutdown     //Active port
Router(config-if)#exit     //Exit port
Router(config)#     //Back to global configuration mode
```

Step 5: Testing

First, test whether the host can ping its own gateway. From Laptop1, we ping the router's Fa0/0 port address (192.168.10.1). The results are shown in Figure 2.21.

Figure 2.21　Results of Laptop1 ping router's Fa0/0 port

From Laptop3, we can ping the router's Fa0/1 port address (192.168.20.1). The results are shown in Figure 2.22.

Figure 2.22　Results of Laptop3 ping the Fa0/1 port of the router

Clearly, hosts from both Network A and Network B can communicate with their respective gateways, as shown in Figure 2.23.

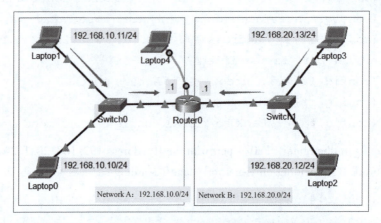

Figure 2.23　Schematic diagram of communication between hosts and their own gateway within the network

Based on our understanding, Router0 should be able to forward data between different networks, indicating that Network A and Network B should be able to communicate normally. We now test this by pinging Laptop3 from Laptop1 to see if data can be transferred from Network A to Network B, with the results shown in Figure 2.24.

```
C:\>ping 192.168.20.13

Pinging 192.168.20.13 with 32 bytes of data:

Request timed out.
Request timed out.
Request timed out.
Request timed out.

Ping statistics for 192.168.20.13:
    Packets: Sent = 4, Received = 0, Lost = 4 (100% loss),
```

Figure 2.24　Results of ping Laptop3 from Laptop1

It still results in "Request timed out." Why can't Laptop1 and Laptop3 from different networks communicate?

In this case, Laptop1 prepares to send data and must determine the recipient's network. If it's the same network, it sends directly; if not, it sends to the gateway for forwarding. Essentially, like a person sending a parcel: if the recipient is on the same street, they deliver directly; otherwise, they use a courier. Here, no one has informed Laptop1 of its gateway's identity or IP, akin to not knowing which courier to use for delivery. Thus, the solution is to inform all network hosts of their gateways, essentially the IP of the gateway.

4. Configure gateway parameters for the host

How do we configure the gateway parameter for the host? In the IP configuration interface, as shown in Figure 2.25, we set Laptop1's gateway information to 192.168.10.1, informing it to send data to this address for forwarding to other networks.

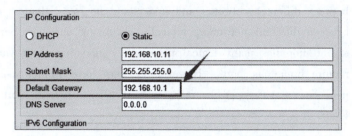

Figure 2.25 The configuration gateway for Laptop1

The same method is applied to configure the gateway for other hosts. Note: The gateway for Network B is 192.168.20.1.

Finally, we test the inter-network communication by pinging Laptop3 from Laptop1, with the results shown in Figure 2.26.

```
C:\>ping 192.168.20.13

Pinging 192.168.20.13 with 32 bytes of data:

Reply from 192.168.20.13: bytes=32 time=1ms TTL=127
Reply from 192.168.20.13: bytes=32 time=1ms TTL=127
Reply from 192.168.20.13: bytes=32 time<1ms TTL=127
Reply from 192.168.20.13: bytes=32 time<1ms TTL=127

Ping statistics for 192.168.20.13:
    Packets: Sent = 4, Received = 4, Lost = 0 (0% loss),
Approximate round trip times in milli-seconds:
    Minimum = 0ms, Maximum = 1ms, Average = 0ms
```

Figure 2.26 Results of ping Laptop3 from Laptop1

Laptop1 and Laptop3 can communicate normally, indicating that cross-network data transmission has been successfully implemented.

Exercise: Based on the topology from Task 2, expand two networks. Following this task, connect the two different networks by using a router, configure appropriately, and ensure all network users can ping each other.

Introduction of ideological and political education: A gateway is a connection point between different networks, which is like the international border or port. It facilitates data communication between networks and often houses security policies to filter hazardous data, ensuring the safe transmission of information. In China, gateways to the Internet also play a critical role in maintaining cyberspace security and national sovereignty. Committed to peaceful development, China emphasizes mutual security and the building of a community with a shared future for mankind, promoting a world oriented towards mutual benefits and collective safety.

Task 4: Connecting Different Networks by Using Layer 3 Switches—SVI (VLAN, Trunk)

Connecting Different Networks by Using Layer 3 Switches—SVI (VLAN,Trunk)

Task Objective: Utilize a Layer 3 switches to facilitate communication between different networks and understand its role. Routers can forward data between different networks, typically occupying one port per network. If an organization has many logical networks, connecting them via a router would require numerous ports. However, routers have limited port numbers. For instance, a Cisco 2811 router only has two default ports, as shown in Figure 2.27.

Figure 2.27 Cisco 2811 Router Backplane Interface

Indeed, we can purchase more modules to insert into expansion slots for additional network ports, but this increases deployment costs, making it less preferable.

A common solution is to use a Layer 3 switches instead of a router for forwarding data between different networks.

1. Connect Different Networks by Using Layer 3 Switches

When replacing the router in Project Task 3 with a Cisco 3,560 Layer 3 switches, as shown in Figure 2.28, the Fa0/1 port of the Layer 3 switches, MS0, can serve as the gateway for Network A, and the Fa0/2 port as the gateway for Network B.

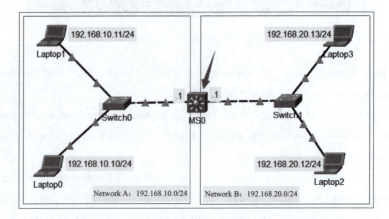

Figure 2.28 Deploying Cisco 3560 layer 3 switches

2. Configure the Layer 3 Switches to Enable Host Communication Between Different Networks

IP configurations are shown in Table 2.8.

Table 2.8 Router Router0 Port IP Address Allocation Table

Port name	IP address	Notes
Fa0/1	192.168.10.1/24	Must be on the same network as Laptop0 and 1
Fa0/2	192.168.20.1/24	Must be on the same network as Laptop2 and 3

Similarly, connect to MS0's console port and log into the console interface by using the Terminal tool to configure as follows:

```
Switch>enable      //Enter privileged mode
Switch#configure terminal    //Enter the global configuration mode
Switch(config)# hostname MS0   //Set the switch name to MS0 (Not necessary, no impact on communication, just easy to manage, but it's a good habit)
MS0(config)#interface fa0/1   //The Fa0/1 port is displayed
MS0(config-if)#no switchport   //Disable the switching feature of ports (By default, switch ports cannot be configured with IP addresses)
MS0(config-if)#ip address 192.168.10.1 255.255.255.0   //Configure an IP address as the gateway of network A
MS0(config-if)#exit    //Exit the Fa0/1 port
MS0(config)#interface fa0/2    //Enter Fa0/2 port
MS0(config-if)#no switchport    //Turn off switching features of a port
MS0(config-if)#ip address 192.168.20.1 255.255.255.0   //Configure the IP address as the gateway for Network B
MS0(config-if)#exit    //Exit Fa0/2 port
MS0(config)#ip routing    //Enable IP routing function (Cisco 3560 layer 3 switch has routing function, but it is turned off by default and needs to be enabled manually)
```

Ping from Laptop1 to Laptop3 to test if the cross-network communication is functioning normally. The results are shown in Figure 2.29, indicating successful communication.

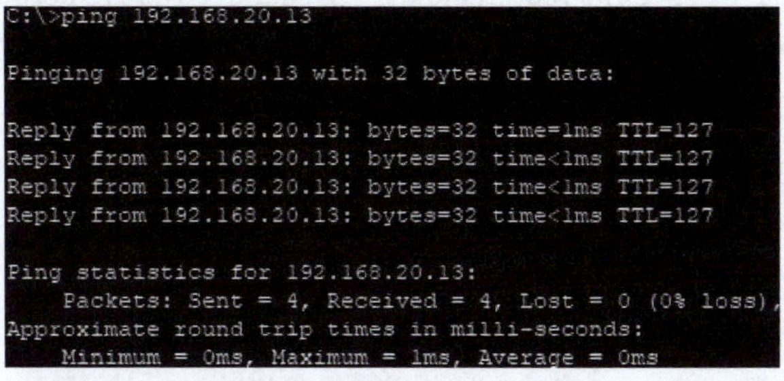

Figure 2.29 Results of ping Laptop3 from Laptop1

However, the method of disabling the switch port's switching functionality to convert it into a Layer 3 routing port isn't universally applicable.

For instance, consider a scenario where Networks A and B represent the logical networks of two departments (Sales and R&D) in a company. As the Sales department grows, all the switch ports get occupied, leaving no ports for new employees. The solution might involve connecting new employees' computers to switches of other departments for network access, as illustrated in Figure 2.30.

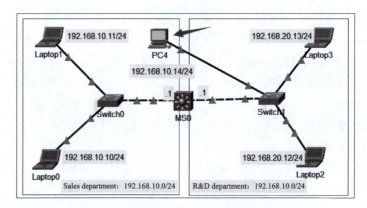

Figure 2.30　Adding a Computer named PC4

When PC4 attempts to communicate with computers in its own department, the data must pass through MS0's Fa0/2 port. However, since this port serves as the gateway for the R&D department, it's in a different network from PC4, leading to a communication failure. This design limitation prevents PC4 from communicating with other computers in its own department.

Is there a way to further optimize the network for scalability? Yes, the use of Switch Virtual Interface (SVI) technology can address this issue.

3. Optimizing the Network with SVI Technology

The solution involves introducing the concept of VLANs, which divides a physical network into multiple logical networks managed as separate units.

A logical network, unlike a physical network, is a grouping based on the physical network but can be divided according to specific requirements. A single physical network can be divided into different logical networks, while the physical network remains unchanged. Logical networks are often divided based on geographical locations or user groups.

VLAN consists of devices and users organized into a logical group, irrespective of their physical locations. They can be organized based on functions, departments, or applications, enabling communication as if they are in the same physical network.

In our case, the entire network is a physical one where all devices are interconnected through hardware and media. However, we've planned two different networks with distinct network addresses based on departments, effectively creating two logical networks, or two VLAN.

Project 2　Expanding the Network

Creating different VLANs for separate departments primarily serves to distinguish and isolate data traffic. First, VLAN switches tag incoming data frames to identify them as part of a specific VLAN, thereby separating traffic of different logical networks. Second, layer 2 switches restrict data transmission within the VLAN, preventing unauthorized access to data and enhancing security.

VLANs also isolate network broadcasts within one LAN. By limiting broadcast packets to a specific VLAN, they protect other VLANs from excessive resource consumption, ensuring stable network efficiency.

Now, let's re-plan and configure our network using VLANs to enable inter-departmental communication. To avoid affecting subsequent configurations, we first clear all previous configurations on the Layer 3 switches MS0, with the following command:

```
    MS0(config)# interface fastEthernet 0/1      //Enter FastEthernet 0/1 port
    MS0(config-if)#no ip address      //Delete the originally configured IP address
    MS0(config-if)#switchport      //Restoring the switching function of the port
    MS0(config-if)#exit     //Exit port
    MS0(config)# interface fastEthernet 0/2      //Enter FastEthernet 0/2 port
    MS0(config-if)#no ip address      //Delete the originally configured IP address
    MS0(config-if)#switchport      //Restoring the switching function of the port
    MS0(config-if)#exit     //Exit port
```

Then, we begin implementing this task.

Step 1: Plan VLANs for the two departments, as shown in Table 2.9.

Table 2.9　VLAN Plan

Department Name	VLAN number	VLAN name	Default gateway	Switch port
Sales	10	Sales	192.168.10.1/24	Switch0 的 Fa0/1、Fa0/4; Switch1 的 Fa0/1
Development	20	Development	192.168.20.1/24	Switch1 的 Fa0/12、Fa0/3

Step 2: As per the planning, create VLAN10 for the sales department and assign the switch ports connected to the sales department computers to VLAN 10.

On Switch0, create VLAN 10 and assign ports Fa0/1 and Fa0/4, used by the sales department computers, to VLAN 10 with the following command:

```
Switch>
Switch>enable     //Enter privileged mode
Switch#configure terminal     //Enter the global configuration mode
Switch(config)#hostname Switch0     //Modify the host name (to facilitate management)
Switch0(config)#vlan 10     //Create VLAN 10
Switch0(config-vlan)#name Sales  //Define a name for VLAN 10: Sales
Switch0(config-vlan)#exit    //Exit the configuration of VLAN 10
Switch0(config)#interface fa0/1     //The Fa0/1 port is displayed
Switch0(config-if)#switchport mode access     //Change the working mode of the port to user access mode
Switch0(config-if)#switchport access vlan 10    //Add the port to VLAN 10
Switch0(config-if)#exit
Switch0(config)#interface fa0/4     //Enter fa0/4 port
Switch0(config-if)#switchport mode access     //Change the working mode of the port to
```

Create VLAN 10 on Switch1 and transfer the port Fa0/1 occupied by the sales department computer to VLAN 10. The command is as follows:

```
Switch>
Switch>enable     //Enter privileged mode
Switch#configure terminal     //Enter the global configuration mode
Switch(config)#hostname Switch1     //Modify the host name (to facilitate management)
Switch1(config)#vlan 10     //Create VLAN 10
Switch1(config-vlan)#name Sales  //Define a name for VLAN 10: Sales
Switch1(config-vlan)#exit    //Exit the configuration of VLAN 10
Switch1(config)#interface fa0/1     //The Fa0/1 port is displayed
Switch1(config-if)#switchport mode access     //Change the working mode of the port to user access mode
Switch1(config-if)#switchport access vlan 10    //Assign the port to VLAN 10
Switch1(config-if)#exit     //Exit the Fa0/1 port
Switch1(config)#
```

Step 3: Following the plan, create VLAN 20 for the R&D department and assign the switch ports connected to the R&D department computers to VLAN 20.

On Switch1, create VLAN 20 and include the ports Fa0/12 and Fa0/13, used by R&D department computers, into VLAN 20 with the following command:

```
    Switch1(config)#vlan 20        //Create VLAN 20
    Switch1(config-vlan)#name Development      //Define a name for VLAN
20: Development
    Switch1(config-vlan)#exit      //Exit the configuration of VLAN 20
    Switch1(config)#interface range fa0/12-13   //Enter Fa0/12, Fa0/13
port group
    Switch1(config-if)#switchport mode access     //Change the working
mode of the port to user access mode
    Switch1(config-if)#switchport access vlan 20    //Assign the port
to VLAN 20
    Switch1(config-if)#exit     //Exit the Fa0/1 port
    Switch1(config)#
```

Check the configuration of VLAN:

View the VLAN database information in Switch0, as shown in Figure 2.31.

```
Switch0#show vlan

VLAN Name                         Status    Ports
---- -------------------------- --------- -------------------------------
1    default                     active    Fa0/2, Fa0/3, Fa0/5, Fa0/6
                                           Fa0/7, Fa0/8, Fa0/9, Fa0/10
                                           Fa0/11, Fa0/12, Fa0/13, Fa0/14
                                           Fa0/15, Fa0/16, Fa0/17, Fa0/18
                                           Fa0/19, Fa0/20, Fa0/21, Fa0/22
                                           Fa0/23, Gig0/1, Gig0/2
10   Sales                       active    Fa0/1, Fa0/4
1002 fddi-default                active
1003 token-ring-default          active
1004 fddinet-default             active
1005 trnet-default               active
```

Figure 2.31   VLAN database information for Switch0

View the VLAN database information in Switch1, as shown in Figure 2.32.

```
Switch1#show vlan

VLAN Name                         Status    Ports
---- -------------------------- --------- -------------------------------
1    default                     active    Fa0/2, Fa0/3, Fa0/4, Fa0/5
                                           Fa0/6, Fa0/7, Fa0/8, Fa0/9
                                           Fa0/10, Fa0/11, Fa0/14, Fa0/15
                                           Fa0/16, Fa0/17, Fa0/18, Fa0/19
                                           Fa0/20, Fa0/21, Fa0/22, Fa0/23
                                           Gig0/1, Gig0/2
10   Sales                       active    Fa0/1
20   Development                 active    Fa0/12, Fa0/13
1002 fddi-default                active
1003 token-ring-default          active
1004 fddinet-default             active
1005 trnet-default               active
```

Figure 2.32   VLAN database information for Switch1

Upon inspection, it's found that all VLAN 10 and 20 are created, and the user-occupied ports are assigned to the respective VLAN.

Step 4: Since the link between switches carries data for all VLANs and isn't dedicated to a specific VLAN or user access, the interconnecting ports shouldn't be set to the access mode.

Instead, the link between Switch0 and MS0 should be set as a trunk.

To configure trunking, set Switch0's Fa0/24 port and MS0's Fa0/1 port to the trunk mode with the following commands, starting with the Layer 2 switches which is named as Switch0.

```
Switch0(config)# interface fastEthernet 0/24    //Enter FastEthernet port 0/24
Switch0(config-if)#switchport mode trunk    //Set the port working mode to trunk mode
Switch0(config-if)#exit    //Exit port
Switch0(config)#
```

Configure Core Switch MS0:

```
MS0(config)# interface fastEthernet 0/1    //Enter FastEthernet 0/1 port
MS0(config-if)#switchport trunk encapsulation dot1q    //Define the encapsulation protocol as dot1q
MS0(config-if)#switchport mode trunk    //Set the port working mode to trunk mode
MS0(config-if)#exit    //Exit port
MS0(config)#
```

Step 5: Set the link between Switch1 and MS0 as a trunk. Configure the Switch1's Fa0/24 port and MS0's Fa0/2 port to the trunk mode with the following commands:

Configure Layer 2 Switches Switch1:

```
Switch1(config)# interface fastEthernet 0/24    //Enter FastEthernet port 0/24
Switch1(config-if)#switchport mode trunk    //Set the port working mode to trunk mode
Switch1(config-if)#exit    //Exit port
Switch1(config)#
```

Configure Core Switch MS0:

```
MS0(config)# interface fastEthernet 0/2    //Enter FastEthernet 0/2 port
MS0(config-if)#switchport trunk encapsulation dot1q    //Define the encapsulation protocol as dot1q
MS0(config-if)#switchport mode trunk    //Set the port working mode to trunk mode
MS0(config-if)#exit    //Exit port
MS0(config)#
```

Check the trunk port status of switch Swith0, as shown in Figure 2.33.

```
Switch0#show interfaces trunk
Port        Mode        Encapsulation   Status      Native vlan
Fa0/24      on          802.1q          trunking    1
```

Figure 2.33　Trunk Port Status of Swith0

If the mode of the switch interconnect port Fa0/24 is "on" and the status is "truncating", it indicates that the port status is normal.

Check the trunk port status of other switches, as shown in Figures 2.34 and 2.35.

```
Switch1#show interfaces trunk
Port        Mode        Encapsulation   Status      Native vlan
Fa0/24      on          802.1q          trunking    1
```

Figure 2.34　Trunk Port Status of Swith1

```
MS0#show interfaces trunk
Port        Mode        Encapsulation   Status      Native vlan
Fa0/1       on          802.1q          trunking    1
Fa0/2       on          802.1q          trunking    1
```

Figure 2.35　Trunk Port Status of MS0

The above information indicates that the status of all trunk ports on switches Switch1 and MS0 is normal.

Step 6: Create VLANs 10 and 20 on MS0 to carry the data of VLAN 10 and 20.

```
MS0(config)#vlan 10         //Create VLAN 10
MS0(config-vlan)#name Sales      //Give Vlan 10 a name: Sales
MS0(config-vlan)#exit       //Exit the configuration of VLAN 10
MS0(config)#vlan 20         //Create VLAN 20
MS0(config-vlan)#name Development //Give Vlan 20 a name: Development
MS0(config-vlan)#exit       //Exit VLAN 20 configuration
```

Check the VLAN database information on MS0, and the results are shown in Figure 2.36.

```
MS0#show vlan

VLAN Name                             Status    Ports
---- -------------------------------- --------- -------------------------------
1    default                          active    Fa0/3, Fa0/4, Fa0/5, Fa0/6
                                                Fa0/7, Fa0/8, Fa0/9, Fa0/10
                                                Fa0/11, Fa0/12, Fa0/13, Fa0/14
                                                Fa0/15, Fa0/16, Fa0/17, Fa0/18
                                                Fa0/19, Fa0/20, Fa0/21, Fa0/22
                                                Fa0/23, Fa0/24, Gig0/1, Gig0/2
10   Sales                            active
20   Development                      active
1002 fddi-default                     active
1003 token-ring-default               active
1004 fddinet-default                  active
1005 trnet-default                    active
```

Figure 2.36　VLAN Database Information for MS0

VLAN 10 and 20 have been successfully created, and the names of the VLANs are consistent. (Note: The names of the VLANs must match; otherwise, network devices will not recognize them as the same VLAN, even if the VLAN number is the same).

Now, PC4 should be able to communicate across switches with users in the same department. The test results are shown in Figure 2.37.

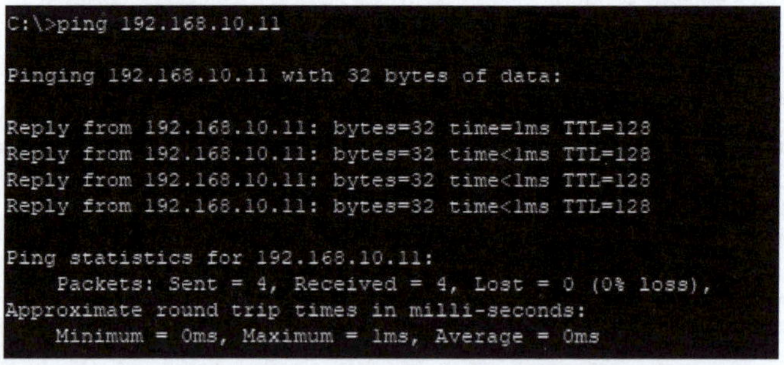

Figure 2.37 PC4 ping Laptop1 result

Step 7: Configure IP addresses for VLAN 10 and 20 ports on MS0 to provide gateway services to users in VLAN 10 and 20, enabling data forwarding between the sales and R&D departments.

The command in MS0 is as follows:

```
    MS0(config)#interface vlan 10    //Enter vlan10 port (note, it is a port)
    MS0(config-if)#ip address 192.168.10.1 255.255.255.0    //Configure an IP address to use as the gateway for VLAN 10
    MS0(config-if)#exit    //Exit
    MS0(config)#interface vlan 20    //Enter vlan20 port (note, it is a port)
    MS0(config-if)#ip address 192.168.20.1 255.255.255.0    //Configure an IP address to use as the gateway for VLAN 20
    MS0(config-if)#exit    //Exit
    MS0(config)#
```

Users from both departments should be able to ping their respective gateways and communicate normally. The results are shown in Figures 2.38 and 2.39.

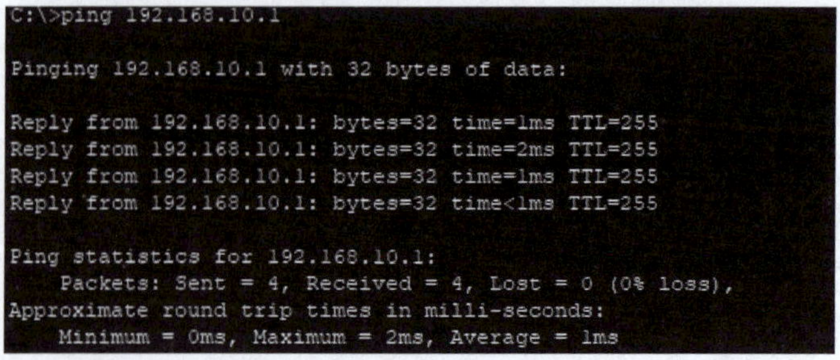

Figure 2.38 Results of Laptop1 ping its own gateway

```
C:\>ping 192.168.20.1

Pinging 192.168.20.1 with 32 bytes of data:

Reply from 192.168.20.1: bytes=32 time=1ms TTL=255
Reply from 192.168.20.1: bytes=32 time<1ms TTL=255
Reply from 192.168.20.1: bytes=32 time<1ms TTL=255
Reply from 192.168.20.1: bytes=32 time<1ms TTL=255

Ping statistics for 192.168.20.1:
    Packets: Sent = 4, Received = 4, Lost = 0 (0% loss),
Approximate round trip times in milli-seconds:
    Minimum = 0ms, Maximum = 1ms, Average = 0ms
```

Figure 2.39 Results of Laptop3 ping its own gateway

Step 8: Test communication between different networks. Ping Laptop3 in the R&D department from Laptop1 in the Sales department. The results are shown in Figure 2.40.

```
C:\>ping 192.168.20.13

Pinging 192.168.20.13 with 32 bytes of data:

Reply from 192.168.20.13: bytes=32 time=1ms TTL=127
Reply from 192.168.20.13: bytes=32 time<1ms TTL=127
Reply from 192.168.20.13: bytes=32 time<1ms TTL=127
Reply from 192.168.20.13: bytes=32 time<1ms TTL=127

Ping statistics for 192.168.20.13:
    Packets: Sent = 4, Received = 4, Lost = 0 (0% loss),
Approximate round trip times in milli-seconds:
    Minimum = 0ms, Maximum = 1ms, Average = 0ms
```

Figure 2.40 Results of Laptop1 ping Laptop3

The technologies of VLAN, trunk, and SVI involved in this task, though basic, are among the most practical and crucial ones due to their widespread use in nearly every network. Mastering these technologies will significantly help in further learning.

Classroom Exercise: Based on the topology from Task 3 of this project, replace the router with a Layer 3 switches. Use SVI as the gateway for users on the Layer 3 switches, configure appropriately to achieve full network intercommunication.

Introduction of ideological and political education: SVI and VLAN represent a virtualization approach that addresses network scalability issues. When traditional technologies hit limitations, it's crucial to creatively propose new solutions. Similarly, as we encounter various challenges in life, we must learn to respond proactively, continually enhancing our sense of duty and mission to drive technological progress through reform and innovation, and commit ourselves with urgency to the practice of reform and innovation, contributing to society and realizing personal value.

Task 5: Deploy Wireless Local Area Network—WLAN

Task Objective: Master the basic WLAN technology to provide network access services for mobile users.

Deploy Wireless Local Area Network—WLAN

There's a joke about archaeologists from various countries who, during a collaborative project, shared their findings. An Egyptian archaeologist claimed they had found coaxial cables in the pyramids, suggesting their ancestors might have used cables for data transmission. A British archaeologist boasted about finding optical fibers in ancient castles, indicating early communication via fiber optics. A Chinese archaeologist calmly stated they found nothing, implying their ancestors might have always used wireless communication.

Wireless transmission marks a milestone in human communication history. It uses radio waves or electric and magnetic fields instead of wires or cables for data transmission, offering significant convenience for users' accesses.

The simplest way to set up a WLAN is by deploying a wireless Access Point (AP), to which users' computers connect via wireless network cards. The location of the wireless AP in Cisco Packet Tracer is shown in Figure 2.41.

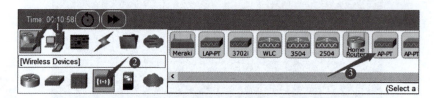

Figure 2.41 Location of Wireless APs in Cisco Packet Tracer

Let's learn about WLAN deployment through a case study. As the Sales department grows, the existing switch ports are insufficient. The network administrator plans to deploy an AP for wireless network accesses for the Sales staff.

Step 1: Deploy the wireless AP in the network as shown in topology Figure 2.42.

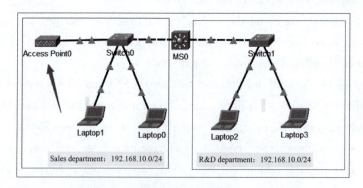

Figure 2.42 Deploying Wireless AP

Step 2: Install a wireless card in PC4, enabling it to access the network via the wireless AP. First, remove the twisted pair cable connected to PC4 and power it down as shown in Figure 2.43.

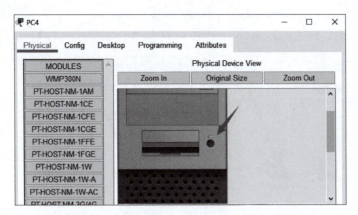

Figure 2.43　The location of host power

Drag the network card to "MODULES" and remove the original RJ45 network card, as shown in Figure 2.44.

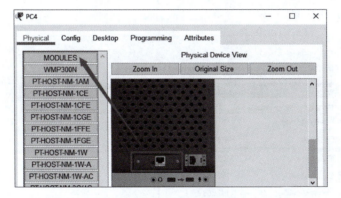

Figure 2.44　The location of host network card

Drag the wireless network card with model "WMP300N" to the network card slot in MODULES, as shown in Figure 2.45.

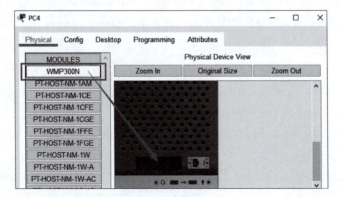

Figure 2.45　WMP300N Installing Wireless Network Card WMP300N

Note: After installing the wireless network card, remember to power on PC4 (which was previously turned off). PC4 will connect to the wireless AP, as shown in Figure 2.46.

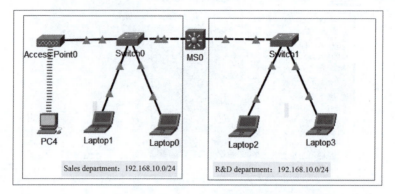

Figure 2.46　Wireless Connection Successful

Step 3: Configure PC4 with an IP address (192.168.10.14/24) and a gateway (192.168.10.1), as illustrated in Figure 2.47.

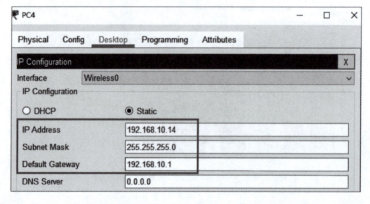

Figure 2.47　Configuring the IP address of the PC4 wireless network card

Step 4: On Switch0, assign the port (Fa0/23) where the wireless AP is connected to the Virtual Local Area Network (VLAN 10) of the sales department, as per the following command:

```
Switch0>enable     //Enter privileged mode
Switch0#configure terminal    //Enter the global configuration mode
Switch0(config)#interface fastEthernet 0/23     //Enter the port occupied by the wireless AP (Fa0/23)
Switch0(config-if)#switchport mode access     //Set the access mode of the port
Switch0(config-if)#switchport access vlan 10     //Add the port to VLAN 10
Switch0(config-if)#exit
Switch0(config)#
```

Check Switch0's VLAN database to confirm the successful assignment; the result is shown in Figure 2.48.

```
Switch0#show vlan

VLAN Name                             Status    Ports
---- -------------------------------- --------- -------------------------------
1    default                          active    Fa0/2, Fa0/3, Fa0/5, Fa0/6
                                                Fa0/7, Fa0/8, Fa0/9, Fa0/10
                                                Fa0/11, Fa0/12, Fa0/13, Fa0/14
                                                Fa0/15, Fa0/16, Fa0/17, Fa0/18
                                                Fa0/19, Fa0/20, Fa0/21, Fa0/22
                                                Gig0/1, Gig0/2
10   Sales                            active    Fa0/1, Fa0/4, Fa0/23
1002 fddi-default                     active
1003 token-ring-default               active
1004 fddinet-default                  active
1005 trnet-default                    active
```

Figure 2.48　VLAN database information of Switch0

Fa0/23 is confirmed to be successfully assigned to VLAN 10.

Step 5: Test connectivity by pinging Laptop2, a host in the research department, from PC4; the result is illustrated in Figure 2.49.

```
C:\>ping 192.168.20.12

Pinging 192.168.20.12 with 32 bytes of data:

Reply from 192.168.20.12: bytes=32 time=41ms TTL=127
Reply from 192.168.20.12: bytes=32 time=27ms TTL=127
Reply from 192.168.20.12: bytes=32 time=20ms TTL=127
Reply from 192.168.20.12: bytes=32 time=13ms TTL=127

Ping statistics for 192.168.20.12:
    Packets: Sent = 4, Received = 4, Lost = 0 (0% loss),
Approximate round trip times in milli-seconds:
    Minimum = 13ms, Maximum = 41ms, Average = 25ms
```

Figure 2.49　The result of pinging Laptop2 from PC4

The test results indicate that the communication between PC4 and Laptop2 is normal, confirming that PC4 can successfully access the LAN wirelessly and maintain robust data communication.

This introduction merely showcases an alternative network access method—wireless. In reality, most modern laptops and some desktops come equipped with wireless cards, eliminating the need for additional purchases. Moreover, wireless AP requires more configurations for secure accesses, but due to significant operational differences between Cisco's simulator and actual devices, further demonstrations are limited here.

Classroom exercise: Deploy a wireless AP in the network to enable users to access the network wirelessly.

Literature expansion: The use of wireless technology for data transmission is undoubtedly a revolution for human society, and it has once again liberated mankind. But things are always contradictory, and the inherent flaws of wireless technology also affect the security of data transmission. It is necessary to improve information security awareness and prevent the leakage of sensitive information when transmitting data through wireless networks. We cannot use the security flaws of wireless technology to illegally steal data. Any organization or individual must operate within the scope of the Constitution and law, and all illegal activities should be prosecuted by law.

Summary and Expansion

1. Applications of Switches, Routers, Wireless APs

If the sender and the receiver of a communication are in the same logical network (same network address), a switch is used for interconnection. If they are in different logical networks (different network addresses), a layer 3 devices (layer 3 switches or router) is used as an intermediary for data forwarding. Wireless APs are used to provide WiFi access for mobile users.

2. Differences Among the Three Common Modes of Cisco Network Devices

Cisco routers and switches have three command line modes: user's mode, privileged mode, and global configuration mode (other sub-configuration modes exist but are not detailed here).

These modes differ in terms of the permissions and functions they offer.

The basic user's mode has the least privileges and is the default mode on device startup, allowing access to basic information and some network testing tools like ping and traceroute.

The privileged mode, accessed by the "enable" command from the user mode, offers the highest level of device control.

The global configuration mode, entered via the "configure terminal" command in the privileged mode, is designed for configuring the device, with most configuration tasks executed here or in its sub-modes.

Switching between different modes is often necessary to complete tasks. For example, to configure a device, one must transit from the user mode to the privileged mode, then to the global configuration mode. This can be linked to moving through different rooms in a house for specific functions, like entering the living room before going to the bedroom to sleep, instead of climbing directly through the bedroom window.

3. Practical Commands and Tips for Cisco Network Devices

During configuration, the "?" and "Tab" keys can assist with command query or completion.

The question mark ("?") can be used to inquire about commands supported in the current mode. For example, to find out what commands can be executed in the privileged mode, you can type "?" in that mode, as shown in Figure 2.50.

Each executable command is also annotated with its function. You can learn about the use of commands by consulting the technical documentation about the product provided by the manufacturer. You can also query commands that start with certain known strings. For instance, if you want to query commands starting with "dis", the operation is shown in Figure 2.51.

According to the retrieved information, there are two commands starting with "dis" in that mode: disable and disconnect.

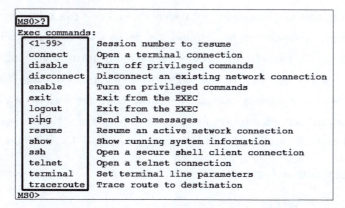

```
MS0>?
Exec commands:
  <1-99>      Session number to resume
  connect     Open a terminal connection
  disable     Turn off privileged commands
  disconnect  Disconnect an existing network connection
  enable      Turn on privileged commands
  exit        Exit from the EXEC
  logout      Exit from the EXEC
  ping        Send echo messages
  resume      Resume an active network connection
  show        Show running system information
  ssh         Open a secure shell client connection
  telnet      Open a telnet connection
  terminal    Set terminal line parameters
  traceroute  Trace route to destination
MS0>
```

Figure 2.50 Use "?" to query executable commands

```
MS0#
MS0#dis?
disable   disconnect
MS0#dis
```

Figure 2.51 Query commands starting with "dis"

The Tab key helps engineers autocomplete commands when the entered command string is unique. What does unique mean? As in the case of Figure 2.51, there are two commands starting with "dis," which is not unique, so pressing Tab won't autocomplete the command because the device doesn't know whether you want to complete "disable" or "disconnect." However, if you enter "disa" and then press the Tab key, the system automatically completes the command, displaying the full "disable"; if you enter "disc" and then press Tab, the system autocompletes it to show the full "disconnect," as illustrated in Figure 2.52.

In fact, when the command string entered is unique, even if the command is incomplete, it can still be executed. Because the system identifies that there is only one command starting with that particular string, it is unique. Once familiar, in most cases, this is how we operate; it is not necessary to type out every command in full, as this approach increases efficiency. The command is as follows:

```
MS0>disc
MS0>disconnect
```

Figure 2.52 Using the Tab key to complete the command

```
MS0>
MS0>en      //abbreviation for enable
MS0#conf t  //abbreviation for configure terminal
MS0(config)#int vlan 10  //abbreviation for interface
MS0(config-if)#ip add 192.168.10.1 255.255.255.0  //abbreviation for ip address
MS0(config-if)#no sh  //abbreviation for no shutdown
MS0(config-if)#ex  //abbreviation for exit
MS0(config)#ex  //abbreviation for exit
MS0#
```

It is important to note that commands are merely tools to assist in achieving objectives; rather than memorizing them, understanding the intricacies of network data communication and fostering a problem-solving mindset is crucial. The society values those artists who, with imagination, can express ideas into creations by using various tools, over who are merely knowledgeable about the tools themselves. As Zhang Sanfeng said about martial arts, the highest level is forgetting the moves; similarly, a great network engineer focuses on problems and solutions, not commands. Even Einstein admitted he could not memorize all his formulas but knew he could look them up.

4. An Analogy about Gateways-Border

To transfer data from Network A to Network B, it first enters through Network A's gateway, then is forwarded by Network A's gateway to Network B's gateway (if Networks A and B are adjacent), and finally exits from Network B's gateway to be delivered to the recipient within Network B, as illustrated in Figure 2.53.

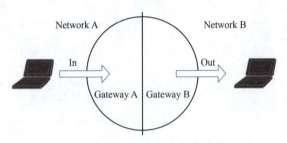

Figure 2.53　Gateway Diagram

In the Internet, "gateways" can be informally linked to border crossings between countries. As depicted in Figure 2.54, with a river demarcating the boundary, Chu and Han have two ports: AC and BD. A and B are Chu's borders, while C and D are Han's. Chu people who want to enter Han must transit through these ports, crossing the river from their own border and entering through Han's border. Any attempt to cross the border irregularly without following the official routes, is considered illegal and not permitted.

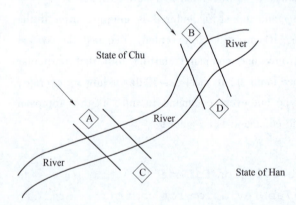

Figure 2.54　The Borde of Chu and Han

From one network to another, there may be multiple channels for data transfer. Similarly, like multiple ports between Chu and Han, a network can have various gateways.

The gateway IP address configured in a user's computer determines where data is sent for forwarding to another network. Just as a person from Chu decides whether to enter Han through the AC or BD port based on their knowledge, data routing depends on the configured gateway.

If someone from Chu knows both ports, they might alternate between them for load balancing. However, if they're unaware of any port, they can't leave Chu, just as a computer without a gateway IP cannot find a node to forward its data, preventing it from reaching its destination.

5. The Mechanism of Switch Forwarding Data-MAC Address

Switches use a MAC (Media Access Control) address table stored in their cache to determine through which port data should be forwarded. This can be viewed by using the show mac-address-table command, as shown in Figure 2.55.

```
Switch1#show mac-address-table
          Mac Address Table
-------------------------------------------

Vlan    Mac Address       Type        Ports
----    -----------       ----        -----

 1      0001.c774.3002    DYNAMIC     Fa0/24
 20     0001.42d1.e839    DYNAMIC     Fa0/12
 20     0060.5c74.2c99    DYNAMIC     Fa0/13
```

Figure 2.55 MAC Address Table Information for Switch1

Taking Laptop2 sending data to Laptop3 as an example of, when Laptop2 sends data to Switch1, the switch acts like a courier, reading the recipient's MAC address from the data frame, as shown in Figure 2.56. This process ensures efficient and accurate data delivery within a network.

```
OSI Model    Inbound PDU Details    Outbou

At Device: Switch1
Source: Laptop2
Destination: 192.168.20.13

In Layers
Layer7
Layer6
Layer5
Layer4
Layer3
Layer 2: Ethernet II Header
0001.42D1.E839 >> 0060.5C74.2C99
Layer 1: Port FastEthernet0/12
```

Figure 2.56 Laptop2 Sends Data Frames for Laptop3

Then we can check the MAC address table, as shown in Figure 2.57.

```
Switch1#show mac-address-table
          Mac Address Table
-------------------------------------------

Vlan    Mac Address       Type        Ports
----    -----------       ----        -----

 1      0001.c774.3002    DYNAMIC     Fa0/24
 20     0001.42d1.e839    DYNAMIC     Fa0/12
 20     0060.5c74.2c99    DYNAMIC     Fa0/13
```

Figure 2.57 MAC Address Table Information of Switch1

Upon discovering that the recipient's device is connected to port Fa0/13 and also belongs to VLAN20, the switch sends the data intended for Laptop3 through port Fa0/13. It's important to note that once the switch sends out the data, its job is considered done. The switch does not concern itself with whether Laptop3 receives the data or even verify the existence of Laptop3. As long as there is a record in the MAC address table, the switch will forward the data through the port indicated in the record.

If the switch cannot find a record of the recipient's MAC address in its table, it handles this by flooding the data frame through all ports, hoping one of them is correct. While this method can cause data flooding in the network and affect communication efficiency, it serves as an effort to forward the data and minimize packet loss.

The MAC address table in a switch is formed through a process of self-learning. This table essentially records which host occupies which port and the VLAN to which this port is assigned. The MAC address table is stored in the switch's cache memory, meaning it starts empty when the switch is first powered on. The switch actively listens on each port and captures data as it comes in. From this data, the switch learns the MAC address of the "sender" and associates it with the corresponding port number in the MAC address table. This process establishes that a particular host occupies a specific port.

Since the MAC address table is dynamically generated and stored in cache memory, it is lost when the switch is powered off. Additionally, the table can be manually cleared without turning off the switch by using the clear mac-address-table command.

6. MAC Address

MAC (Media Access Control) Address, also known as a physical address, is a unique identifier assigned to network interfaces for communications on the physical network segment. It's pre-programmed into a network card's EPROM (Erasable Programmable Read-Only Memory) chip by the hardware manufacturer.

A MAC address is 48 bits long (6 bytes) and is typically displayed as 12 hexadecimal digits, like 00-16-EA-AE-3C-40. The first half (00-16-EA) represents the manufacturer's unique identifier assigned by IEEE (Institute of Electrical and Electronics Engineers), while the second half (AE-3C-40) denotes the serial number of the specific network product made by that manufacturer.

MAC addresses are globally unique if not altered by using special methods. Unlike IP addresses which change when a device moves to different networks, MAC addresses are static and remain the same across different networks. In a comparison, a MAC address can be thought of as akin to a person's national ID number which is unique and permanent, while an IP address is similar to a residential address which changes based on locations.

7. Control VLAN Traffic Through TRUNK Link

For Cisco switches, by default, all VLAN traffic is allowed across TRUNK links. To manage the VLAN traffic passing through these TRUNK links and allow only specific VLANs, you can use the following command:

```
Switch1(config)#int fa 0/24      //Enter fastEthernet port 0/24
Switch1(config-if)#switchport mode trunk      //Set the port's working mode to trunk (trunk mode)
Switch1(config-if)#switchport trunk allowed vlan 10,20      //Only allow traffic from VLAN 10 and VLAN 20 to pass
```

You can view the VLANs allowed through a specific TRUNK port by using the "show interfaces trunk" command, as shown in Figure 2.58.

```
Switch1#show interfaces trunk
Port        Mode              Encapsulation  Status        Native vlan
Fa0/24      on                802.1q         trunking      1

Port        Vlans allowed on trunk
Fa0/24      10,20

Port        Vlans allowed and active in management domain
Fa0/24      10,20
```

Figure 2.58 View the Allowed Traffic on the TRUNK Link

It is found that port FastEthernet 0/24 of switch Switch1 only allows traffic from VLAN 10 and VLAN 20. However, to deny certain specific VLAN traffic, the following command can be used:

```
Switch1(config)#int fa 0/24      //Enter fastEthernet port 0/24
Switch1(config-if)#switchport mode trunk      //Set the port's working mode to trunk (trunk mode)
Switch1(config-if)#switchport trunk allowed vlan except 10      //All traffic except VLAN 10 is allowed to pass
```

Once again, by using the command "show interfaces trunk" to check the status of port FastEthernet 0/24 on Switch1, the result is shown in figure 2.59.

```
Switch1#sh interfaces trunk
Port        Mode              Encapsulation  Status        Native vlan
Fa0/24      on                802.1q         trunking      1

Port        Vlans allowed on trunk
Fa0/24      1-9,11-1005

Port        Vlans allowed and active in management domain
Fa0/24      1,20
```

Figure 2.59 View the Allowed Traffic on the TRUNK Link

The displayed result shows that traffic from VLAN 1-99 and VLAN 11-1005 is all permitted to pass through the TRUNK port (FastEthernet 0/24).

8. The Mechanism for Forwarding Data by Layer 3 Devices (Routers or Switches) —IP Routing Table

Layer 3 devices (routers or Layer 3 switches) determine which port to forward data based on the IP routing table stored in memory. The command "show ip route" can be used to view the routing table, as illustrated in Figure 2.60.

```
MS0#show ip route
Codes: C - connected, S - static, I - IGRP, R - RIP, M - mobile, B - BGP
       D - EIGRP, EX - EIGRP external, O - OSPF, IA - OSPF inter area
       N1 - OSPF NSSA external type 1, N2 - OSPF NSSA external type 2
       E1 - OSPF external type 1, E2 - OSPF external type 2, E - EGP
       i - IS-IS, L1 - IS-IS level-1, L2 - IS-IS level-2, ia - IS-IS inter
area
       * - candidate default, U - per-user static route, o - ODR
       P - periodic downloaded static route

Gateway of last resort is not set

C    192.168.10.0/24 is directly connected, Vlan10
C    192.168.20.0/24 is directly connected, Vlan20

MS0#
MS0#   The type of Routers   Destination network address   Forwarding port
```

Figure 2.60 Routing Table for MS0

The routing table contains two entries, which record specific information. Here's an interpretation of the first route entry: The initial field "C" indicates a directly-connected network, meaning these networks are directly connected to this device (other codes and their meanings are explained above). The second field shows the network number (192.168.10.0/24) and indicates that this network is directly connected. The last field specifies that data for the 192.168.10.0/24 network should be forwarded through the Vlan10 port.

For example, if Laptop3 from the R&D department (192.168.20.13/24) sends data to Laptop1 in the Sales department (192.168.10.11/24), Laptop3 sends the data packet to its gateway (vlan20: 192.168.20.1/24) first. After MS0 receives the data through the vlan20 port, it, like a courier, reads the destination IP address (192.168.10.11) on the packet. Then MS0 checks its routing table, where it matches the entry for the network address 192.168.10.0/24 (if there are multiple entries for the 192.168.10.0 network in the routing table, it matches the entry with the longest subnet mask, following the longest match principle). Based on the matched routing entry, the data is sent out through the Vlan10 port.

Like a switch, once a router forwards a data packet, it completes its task. It doesn't concern itself with whether the recipient will receive it or even if the recipient actually exists. The router strictly follows the instructions in the routing table about which port to forward through.

Unlike a switch forwarding data frames, a router consults the routing table. If it doesn't find a route entry for the destination network, it discards the packet. Imagine a courier who is unable to find any information about a package's destination address simply discards it! This may seem absurd, but that's how routers operate, and it's in line with the rules.

Route entries in the routing table can originate from various sources, broadly categorized into three types. One is static routes, manually configured by network administrators, marked with "S" for "static". Another type is dynamic routes, learned by layer 3 devices through routing protocols (like RIP, OSPF, BGP) from neighbors or calculated based on information received from neighbors. The last type, as seen in this example, are routes for networks directly connected to the router, marked with "C." This routing information is stored in the router's cache and doesn't require input from other devices.

9. Default Subnet Masks for Class A, Class B, and Class C IP addresses

When configuring an IP address for a computer, the system automatically assigns a default subnet mask. This default is determined according to Table 2.10.

Table 2.10 Default Subnet Masks for Various IP Addresses

Network	Default subnet mask	Mask length
A	255.0.0.0	/8
B	255.255.0.0	/16
C	255.255.255.0	/24

10. Fundamentals of Wireless Local Area Network (WLAN)

Wireless Local Area Networks (WLANs) operate without the use of wires or transmission cables, typically over short distances of a few dozen meters by utilizing radio waves or electromagnetic fields for data transmission. However, the backbone of a WLAN usually involves wired cables, with users accessing the network through one or more wireless Access Points (APs).

The first version of WLAN, IEEE 802.11, was released in 1997, defining the medium access control layer and the physical layer. It includes two radio frequency methods and an infrared transmission method which operates in the 2.4 GHz ISM band, with a designed total data transmission rate of 2 Mbit/s. Over decades, WLAN technology has been widely applied in business districts, universities, airports, and other public areas requiring wireless networks.

There's a common misconception that WLAN and WiFi are the same. However, they are not identical concepts. WLAN refers to the standard IEEE 802.11, and WiFi is just one implementation of this standard, though it is the most widely used by the general public. Besides WiFi, the IEEE 802.11 standard includes Wireless Gigabit (WiGig) Alliance products. The WiGig Alliance merged into the WiFi Alliance on January 4th, 2013.

Up to now, WiFi has evolved into its sixth generation, as detailed in Table 2.11.

Table 2.11 World Representative for WiFi Development

Generation	Year	Standard	Frequency band of work	Maximum speed
First	1997	Original IEEE 802.11	2.4 GHz	2 Mbit/s
Second	1999	IEEE 802.11b	2.4 GHz	11 Mbit/s

continued

Generation	Year	Standard	Frequency band of work	Maximum speed
Third	1999	IEEE 802.11a	5 GHz	54 Mbit/s
Fourth	2009	IEEE 802.11n	2.4 GHz and 5 GHz	600 Mbit/s
Fifth	2013	IEEE 802.11ac	5 GHz	6.9 Gbit/s
Sixth	2019	IEEE 802.11ax	2.4 GHz and 5 GHz	9.6 Gbit/s

11. Save Device Configuration

It's important to remember to save the configuration settings of devices to prevent data loss. Configuration data is temporarily stored in the running-config file in the device's memory, which will be lost if power is disconnected. There's also a permanent startup-config file stored in the device's flash memory.

In privileged mode, to prevent losing configurations after completing your setup, you should use the command copy running-config startup-config to copy temporary data from running-config to startup-config, or simply use the write command to save the configuration.

12. In Non-Privileged Mode, Enforce Commands in Privileged Mode

In configuration mode, we cannot directly execute privileged mode commands like show or ping. However, we can use the do command to forcefully execute them.

Each mode has its corresponding commands. Often, while configuring a device in the configuration mode, we need to check the setup by using show commands or test connectivity with ping. Exiting from the configuration mode to the privileged mode for this purpose can be inefficient. Instead, we can directly use the do command in the configuration mode to execute privileged mode commands. For instance, to save device configuration in the configuration mode, we can use the do write command.

Thinking and Training

(1) The IP address of a host is 172.16.1.1/28. What are the network address, broadcast address, and available host addresses of the host's network?

(2) Can the IP address 172.16.1.16/28 be assigned to a host? Why or why not? What about the address 172.16.1.15/28?

(3) Are the hosts 211.16.1.1/30 and 211.16.1.5/30 in the same logical network? If they need to communicate with each other, through what kind of device must the data be forwarded?

(4) What is the basis for a switch to forward data? How does it forward? Please describe in your own words.

(5) What is the basis for a router to forward data? How does it forward? Please describe in your own words.

(6) Is the statement "Wireless LAN is the same as WiFi" correct? Why or why not?

(7) Refer to the topology shown in Figure 2.61 and design a network system using Cisco Packet Tracer. Configure it appropriately so that all network users can communicate with each other and access the internal server. ×× represents the last two digits of your students ID. Plan the host and gateway IP addresses based on the given IP address range; decide on port numbers, VLAN numbers, and names yourself.

Key tips:

①Create three VLANs to divide the teaching building, the dormitory building, and the server group into three different logical networks.

②Calculate the range of usable host addresses for each logical network based on the given network address, and allocate IP addresses to each host and gateway in the network accordingly.

③Note that the routing function of the Layer 3 switch MS0 must be manually enabled.

④The core skill is developing problem-solving approaches (Calculate host addresses→Allocate IP addresses→Create VLANs→Assign ports accordingly→Configure trunk links→Configure SVI port addresses as gateways on the core switch→Enable routing function).

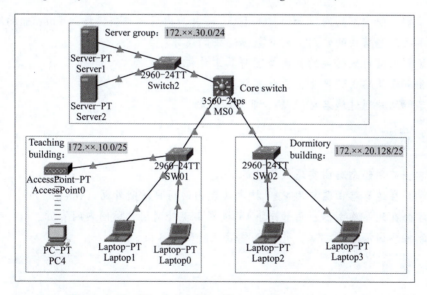

Figure 2.61 Simple topology diagram of campus intranet

项目 3

接入互联网

用户接入网络不只是要访问局域网资源，更多的需求是要访问 Internet 上无穷无尽的资源。本项目的目标是要解决用户计算机访问 Internet 的问题。主要介绍企业网络通过 ISP 的专线接入互联网所需要的核心技术：静态默认路由、地址转换以及域名服务等。

知识目标

（1）了解组织内部的网络如何接入 Internet 并为内网用户提供访问 Internet 的服务；
（2）了解静态路由的作用，培养数据路由的思维；
（3）了解网络地址转换的背景并理解其工作机制；
（4）了解默认路由的产生背景及其作用；
（5）了解用户通过域名访问 Web 页面的过程。

技能目标

（1）学会应用静态路由实现数据包的转发；
（2）学会在边界路由器应用 PAT 技术实现内网用户访问外网（Internet）；
（3）学会在边界路由器应用静态 NAT 技术实现外网用户访问内网资源；
（4）掌握路由汇总的方法，并学会灵活运用默认路由。

相关知识

一般用户接入网络更多的需求是访问 Internet，因为那里的资源几乎无穷无尽。那么用户如何可以访问到 Internet 呢？

是运营商（电信、移动、联通等）为我们提供了访问 Internet 的服务。我们要把数据发到 Internet 上的某台服务器以访问其资源，首先要把这个数据发给某个运营商，然后由运营商通过其错综复杂的网络（中间可能会横跨另外的运营商），将数据发送到目的服务器。

1. 用光纤将企业网络接入运营商的网络

所以我们首先要做的事情就是，选一个你信任的运营商，然后将公司的网络接入运营商的网络，如图 3.1 所示。

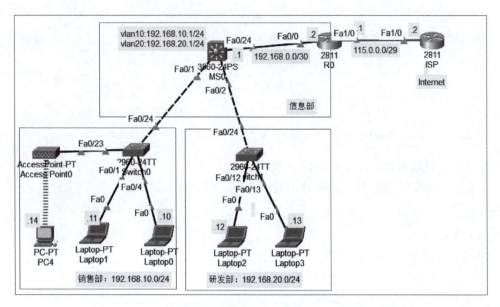

图 3.1　企业网接入运营商网络

公司的内部网络通过路由器 R0 接入运营商的网络。因为这台路由器一边连接的是内网，另一边连接的是外网，这个位置正好是内外网的边界，所以通常称其为边界路由器。

边界路由器一般是通过光纤连接到运营商机房的路由器。边界路由器在公司的机房，距离运营商的机房一般很遥远，所以一般采用光纤接入。这需要边界路由器和运营商的路由器都要安装光模块。在 Cisco Packet Tracer 模拟器中安装光模块的方式如图 3.2 所示。

图 3.2　为路由器安装光模块

光纤在介质列表中的位置如图 3.3 所示。

图 3.3　光纤在介质列表中的位置

2. 配置边界路由器和 ISP 路由器之间的直连链路

核心交换机 MS0 和边界路由器 R0 之间、R0 和 ISP 的路由器之间规划的网络 IP 地址如表 3.1 所示。

表 3.1 IP 地址分配表

设备名称	端口名称	IP 地址	子网掩码	备注
MS0	Fa0/24	192.168.0.1	255.255.255.252	同网
R0	Fa0/0	192.168.0.2	255.255.255.252	
	Fa1/0	115.0.0.1	255.255.255.248	同网
ISP	Fa1/0	115.0.0.2	255.255.255.248	

根据以上规划为对应设备配置 IP 地址，命令如下：

配置三层交换机 MS0 连接边界路由器的端口（Fa0/24）。

```
MS0>en
MS0#conf t
MS0(config)#int fa0/24
MS0(config-if)#no switchport
MS0(config-if)#ip add 192.168.0.1 255.255.255.252
MS0(config-if)#exit
MS0(config)#
```

配置边界路由器 R0 的主机名及其端口。

```
Router>en
Router#conf t
Router(config)#hostname R0
R0(config)#int f 0/0
R0(config-if)#ip add 192.168.0.2 255.255.255.252
R0(config-if)#no sh
R0(config-if)#exit
R0(config)#int fa1/0
R0(config-if)#ip add 115.0.0.1 255.255.255.248
R0(config-if)#no sh
R0(config-if)#exit
R0(config)#
```

配置运营商 ISP 路由器。

```
Router>en
Router#conf t
Router(config)#hostname ISP
ISP(config)#int fa1/0
ISP(config-if)#no sh
ISP(config-if)#ip add 115.0.0.2 255.255.255.248
ISP(config-if)#
```

3. 测试边界路由器和内外网直连设备（MS0 和 ISP 路由器）之间的连通性

测试 MS0→R0→ISP 直连链路的连通性，从 R0 分别 ping 核心交换机 MS0 和 ISP 路由器，结果应该是通的，测试结果如图 3.4 所示。

```
R0#ping 192.168.0.1
Type escape sequence to abort.
Sending 5, 100-byte ICMP Echos to 192.168.0.1, timeout is 2 seconds:
!!!!!
Success rate is 100 percent (5/5), round-trip min/avg/max = 0/0/0 ms
R0#ping 115.0.0.2
Type escape sequence to abort.
Sending 5, 100-byte ICMP Echos to 115.0.0.2, timeout is 2 seconds:
!!!!!
Success rate is 100 percent (5/5), round-trip min/avg/max = 0/2/8 ms
R0#
```

图 3.4 测试结果

在网络设备里执行 ping 命令和 PC 端执行 ping 命令返回的信息是不一样的。路由器用"！"（感叹号）表示通信正常，"."（点号）表示不明原因丢包。通过统计信息"Success rate is 100 percent（5/5）"表示测试了 5 个数据包，成功率 100%。

本项目我们的目标就是让用户能访问到互联网，也就是要让内网用户能 ping 通 ISP 路由器。实现的思路是：首先要让内网用户将数据发到边界路由器，让边界路由器帮忙转发到 ISP；ISP 收到数据包，再给发送方一个确认的应答包，以完成整个通信过程。

所以在这里需要解决的第一个问题就是，企业内网用户如何能将数据发送到边界路由器。如果用户无法将数据发送到边界路由器，那么边界路由器接收不到用户数据，也就无法帮用户将数据包转发到 Internet。接下来我们用静态路由技术来解决这个问题。

任务 1：将数据发送到边界路由器——静态路由

任务目标：实现用户能访问到边界路由器。

这里以 Laptop3 将数据发给边界路由器为例进行分析。Laptop3 封装了一个 IP 数据包，源 IP 地址和目的 IP 地址如表 3.2 所示。

将数据发送到边界路由器——静态路由

表 3.2 Laptop3 发给边界路由器的数据包地址信息

源 IP 地址	目的 IP 地址	目的网络的网络地址
192.168.20.13/24	192.168.0.2/30	192.168.0.0/30

1. 分析内网用户发出 IP 数据包的通信过程

根据本项目的网络拓扑，Laptop3（192.168.20.13/24）和边界路由器（192.168.0.2/30）

是处于不同网络的设备。按照项目 2 的分析，不同网络的设备必须经过网关帮忙转发数据，因此 Laptop3 需要先把数据发送给它自己的网关（192.168.20.1/24），再由三层交换机 MS0 查询路由表，根据路由表的路由条目做出转发决定，丢弃还是转发？从哪个端口转发出去？

项目 2 中我们已经可以确定 Laptop3 可以 ping 通网关，这里不做重复测试。网关拿到 Laptop3 发给边界路由器 R0 的数据以后，为了确定数据如何转发，MS0 需要根据要转发的数据包的目的地址去查路由表。核心路由器 MS0 的路由表如图 3.5 所示。

```
MS0#show ip route
Codes: C - connected, S - static, I - IGRP, R - RIP, M - mobile, B - BGP
       D - EIGRP, EX - EIGRP external, O - OSPF, IA - OSPF inter area
       N1 - OSPF NSSA external type 1, N2 - OSPF NSSA external type 2
       E1 - OSPF external type 1, E2 - OSPF external type 2, E - EGP
       i - IS-IS, L1 - IS-IS level-1, L2 - IS-IS level-2, ia - IS-IS inter
area
       * - candidate default, U - per-user static route, o - ODR
       P - periodic downloaded static route

Gateway of last resort is not set

     192.168.0.0/30 is subnetted, 1 subnets
C       192.168.0.0 is directly connected, FastEthernet0/24
C    192.168.10.0/24 is directly connected, Vlan10
C    192.168.20.0/24 is directly connected, Vlan20
```

图 3.5　核心路由器 MS0 的路由表

在路由表中，MS0 找到目的网络（192.168.0.0）的路由条目。这个路由条目告诉路由器，192.168.0.0 网络就直接连接（directly connected）到了 FastEthernet0/24 端口。如果有数据要发到网络 192.168.0.0，则要将数据从 FastEthernet0/24 端口转发出去。

根据如图 3.1 所示的拓扑，MS0 将数据从 FastEthernet0/24 端口转发出去以后，R0 就可以接收到。整个过程很顺利，不需要任何额外的配置。

2. 分析边界路由器 R0 发应答数据包给内网用户的过程

我们定义一个畅通的信息传递过程是指数据有去有回。目前只能确定 Laptop3 发出的数据可以到达边界路由器 R0，并且边界路由器能接收到该数据包。但是当 R0 要给 Laptop3 发回一个应答包的时候，边界路由器 R0 封装了一个 IP 数据包，其源地址和目的地址如表 3.3 所示。

表 3.3　边界路由器发给 Laptop3 的数据包地址信息

源 IP 地址	目的 IP 地址	目的网络的网络地址
192.168.0.2/30	192.168.20.13/24	192.168.20.0/24

R0 同样也需要根据数据包的目的地址（192.168.20.13/24）查询自己的路由表，才能决定要如何处理该数据包：要丢弃还是转发？从哪个端口将应答包发出？

边界路由器 R0 的路由表如图 3.6 所示。

R0 在路由表中无法查到 Laptop3 所处的网络地址（192.168.20.0/24）信息，按规则只能将应答包丢弃，从而导致通信中断。

有一个简单的办法可以解决，那就是手动往路由表中添加一条 192.168.20.0 网络的记录，这样路由器就可以通过查询路由表作出转发决定。

```
R0#sh ip rou
Codes: C - connected, S - static, I - IGRP, R - RIP, M - mobile, B - BGP
       D - EIGRP, EX - EIGRP external, O - OSPF, IA - OSPF inter area
       N1 - OSPF NSSA external type 1, N2 - OSPF NSSA external type 2
       E1 - OSPF external type 1, E2 - OSPF external type 2, E - EGP
       i - IS-IS, L1 - IS-IS level-1, L2 - IS-IS level-2, ia - IS-IS inter area
       * - candidate default, U - per-user static route, o - ODR
       P - periodic downloaded static route

Gateway of last resort is not set

     115.0.0.0/29 is subnetted, 1 subnets
C       115.0.0.0 is directly connected, FastEthernet1/0
     192.168.0.0/30 is subnetted, 1 subnets
C       192.168.0.0 is directly connected, FastEthernet0/0
R0#
```

图 3.6　边界路由器 R0 的路由表

3. 往路由表中添加静态路由

手动往路由表里添加一条路由信息，这条路由信息就叫静态路由。下面我们将在 R0 的路由表中添加一条信息，告诉路由器如果有发到 192.168.20.0/24 网络的数据，就将数据转发给 MS0（192.168.0.1）。静态路由命令如图 3.7 所示。

```
R0#conf t
R0(config)#ip route 192.168.20.0 255.255.255.0 192.168.0.1
```

图 3.7　静态路由命令

静态路由命令解读：

① "ip route" 命令表示要添加一条静态路由；
② "192.168.20.0 255.255.255.0" 表示最终目的网络的网络号和子网掩码；
③ "192.168.0.1" 表示下一跳地址。

整条命令表达的意思是：如果有数据要发往 192.168.20.0/24 这个网络，则把数据转给 192.168.0.1，由它负责下一步的转发。

检查路由表，关于 192.168.20.0 网络的路由条目，结果如图 3.8 所示。

```
R0#sh ip rou
Codes: C - connected, S - static, I - IGRP, R - RIP, M - mobile, B - BGP
       D - EIGRP, EX - EIGRP external, O - OSPF, IA - OSPF inter area
       N1 - OSPF NSSA external type 1, N2 - OSPF NSSA external type 2
       E1 - OSPF external type 1, E2 - OSPF external type 2, E - EGP
       i - IS-IS, L1 - IS-IS level-1, L2 - IS-IS level-2, ia - IS-IS inter area
       * - candidate default, U - per-user static route, o - ODR
       P - periodic downloaded static route

Gateway of last resort is not set

     115.0.0.0/29 is subnetted, 1 subnets
C       115.0.0.0 is directly connected, FastEthernet1/0
     192.168.0.0/30 is subnetted, 1 subnets
C       192.168.0.0 is directly connected, FastEthernet0/0
S    192.168.20.0/24 [1/0] via 192.168.0.1
```

图 3.8　R0 路由表中新添加的路由条目信息

根据输出的路由表显示，表中多了一条标记 "S" 的 192.168.20.0 的路由。根据上方的解释,"S"表示"static"(静态)。说明去往 192.168.20.0/24 网络的路由条目添加成功。R0 以后查路由表,就可以知道:如果有数据要发送到 192.168.20.0/24 这个网络,就把数据发给 192.168.0.1。

其实 R0 此时并不知道 192.168.0.1 是谁,要从哪个端口转发出去。所以 R0 会再查一次路由表(递归查询),查到记录如图 3.9 所示,才能确定:给 192.168.0.1 的数据是从 FastEthernet0/0 端口发出。

```
R0#sh ip rou
Codes: C - connected, S - static, I - IGRP, R - RIP, M - mobile, B - BGP
       D - EIGRP, EX - EIGRP external, O - OSPF, IA - OSPF inter area
       N1 - OSPF NSSA external type 1, N2 - OSPF NSSA external type 2
       E1 - OSPF external type 1, E2 - OSPF external type 2, E - EGP
       i - IS-IS, L1 - IS-IS level-1, L2 - IS-IS level-2, ia - IS-IS inter area
       * - candidate default, U - per-user static route, o - ODR
       P - periodic downloaded static route

Gateway of last resort is not set

     115.0.0.0/29 is subnetted, 1 subnets
C       115.0.0.0 is directly connected, FastEthernet1/0
     192.168.0.0/30 is subnetted, 1 subnets
C       192.168.0.0 is directly connected, FastEthernet0/0
S    192.168.20.0/24 [1/0] via 192.168.0.1
```

图 3.9　R0 路由表中的路由条目信息

4. 继续分析数据返回的通信过程

数据到达 192.168.0.1(MS0)以后,MS0 也要去查自己的路由表,根据路由表,如图 3.10 所示,查到关于 192.168.20.0/24 网络的信息,最后将数据从 Vlan20 端口转发出去。

```
MS0#show ip route
Codes: C - connected, S - static, I - IGRP, R - RIP, M - mobile, B - BGP
       D - EIGRP, EX - EIGRP external, O - OSPF, IA - OSPF inter area
       N1 - OSPF NSSA external type 1, N2 - OSPF NSSA external type 2
       E1 - OSPF external type 1, E2 - OSPF external type 2, E - EGP
       i - IS-IS, L1 - IS-IS level-1, L2 - IS-IS level-2, ia - IS-IS inter area
       * - candidate default, U - per-user static route, o - ODR
       P - periodic downloaded static route

Gateway of last resort is not set

     192.168.0.0/30 is subnetted, 1 subnets
C       192.168.0.0 is directly connected, FastEthernet0/24
C    192.168.10.0/24 is directly connected, Vlan10
C    192.168.20.0/24 is directly connected, Vlan20
```

图 3.10　MS0 路由表中的路由信息

Laptop3 拿到应答包后,通信完成。Laptop3 主机 ping 边界路由器 R0 的结果如图 3.11 所示。

项目 3　接入互联网

```
C:\>ping 192.168.0.2
Pinging 192.168.0.2 with 32 bytes of data:
Reply from 192.168.0.2: bytes=32 time<1ms TTL=254
Reply from 192.168.0.2: bytes=32 time=3ms TTL=254
Reply from 192.168.0.2: bytes=32 time=7ms TTL=254
Reply from 192.168.0.2: bytes=32 time=1ms TTL=254

Ping statistics for 192.168.0.2:
    Packets: Sent = 4, Received = 4, Lost = 0 (0% loss),
Approximate round trip times in milli-seconds:
    Minimum = 0ms, Maximum = 7ms, Average = 2ms
```

图 3.11　测试结果

5. 在 R0 添加静态路由，实现 VLAN 10 用户能访问到边界路由器

如何实现销售部网络的主机能 ping 通边界路由器？按同样的思维去分析，会发现，数据是可以到达边界路由器 R0 的，只是 R0 不知道应答包该从哪个端口发回来。

可以在 R0 添加一条静态路由，告诉 R0：如果有数据要发到网络 192.168.10.0/24，请转发给 192.168.0.1，命令如下：

```
R0>en
R0#conf t
R0(config)#ip route 192.168.10.0 255.255.255.0 192.168.0.1        //配
置静态路由,发往 192.168.10.0/24 网络的数据,传给节点 192.168.0.1,由它来负责
下一步的转发
```

检查路由表，确认是否添加成功，如图 3.12 所示。

```
R0#sh ip rou
Codes: C - connected, S - static, I - IGRP, R - RIP, M - mobile, B - BGP
       D - EIGRP, EX - EIGRP external, O - OSPF, IA - OSPF inter area
       N1 - OSPF NSSA external type 1, N2 - OSPF NSSA external type 2
       E1 - OSPF external type 1, E2 - OSPF external type 2, E - EGP
       i - IS-IS, L1 - IS-IS level-1, L2 - IS-IS level-2, ia - IS-IS inter area
       * - candidate default, U - per-user static route, o - ODR
       P - periodic downloaded static route

Gateway of last resort is not set

     115.0.0.0/29 is subnetted, 1 subnets
C       115.0.0.0 is directly connected, FastEthernet1/0
     192.168.0.0/30 is subnetted, 1 subnets
C       192.168.0.0 is directly connected, FastEthernet0/0
S    192.168.10.0/24 [1/0] via 192.168.0.1
S    192.168.20.0/24 [1/0] via 192.168.0.1
```

图 3.12　R0 的路由表

确认成功添加路由表以后，测试从 Laptop0 ping 边界路由器（192.168.0.2），结果如图 3.13 所示。

```
C:\>ping 192.168.0.2
Pinging 192.168.0.2 with 32 bytes of data:
Reply from 192.168.0.2: bytes=32 time=1ms TTL=254
Reply from 192.168.0.2: bytes=32 time<1ms TTL=254
Reply from 192.168.0.2: bytes=32 time=1ms TTL=254
Reply from 192.168.0.2: bytes=32 time=1ms TTL=254
```

图 3.13　测试结果

至此，内网用户都可以访问到边界路由器，企业网络内部能全部互通。

掌握了静态路由技术，就可以任意地控制数据的流向。想把数据发给谁，就将静态路由的下一跳指向谁。接下来，如果有数据需要发送到 Internet，则内部主机先将数据发送到边界路由器，再由边界路由器 R0 将数据转发到 ISP 的网络，同样也可以用静态路由来实现将内网数据往 Internet 发送——下一个任务会分析和实现这个目标。

课堂练习：参照本任务，基于企业网络模型，配置静态路由，让所有内网计算机都可以访问到边界路由器 R0。

素质拓展：路由器根据路由表决定如何转发数据包，因此路由表中的路由条目就显得尤为重要。路由表一旦匹配成功，路由器会严格将数据转发给被匹配到的路由条目中记录的下一跳节点。某种程度上说，路由表代表一面旗帜，决定了数据流的方向，所以保持路由表中路由条目的正确性就显得尤为重要。大学生内心也要有一面旗帜，一面爱国主义旗帜，把爱国之情、强国之志、报国之行统一起来，为国家和民族作出应有的贡献。

任务 2：实现内网用户访问互联网服务器——端口地址转换（PAT）

实现内网用户访问互联网服务器——端口地址转换(PAT)

任务目标：在边界路由器转换内网地址，实现内网用户访问外网（ping 通 ISP 路由器地址：115.0.0.2/29）。

经过前面几个任务的配置，企业内网的计算机是否就可以顺利访问到外网了呢？测试发现内网用户还无法访问外网主机。为了找出问题的原因，我们从数据的发出和返回来分析数据通信的整个过程。

继续以本项目的网络拓扑进行分析，Laptop3 要 ping 运营商 115.0.0.2/29 的地址。Laptop3 封装了一个 IP 包，其源地址和目的地址如表 3.4 所示。

表 3.4 Laptop3 发给 ISP 路由器的数据包地址信息

源 IP 地址	目的 IP 地址	目的网络的网络地址
192.168.20.13/24	115.0.0.2/29	115.0.0.0/29

Laptop3 发出的数据包是否能顺利到达 ISP 路由器呢？我们通过分析并且只要在三层设备确保有目的网络的路由条目，数据转发应该就没有问题。

1. Laptop3 发出 IP 数据包后的分析

Laptop3 发现目的地址 115.0.0.2/29 处于不同网络，因此将 IP 数据包发给其网关（MS0：192.168.20.1/24），请求帮忙转发。MS0 收到该 IP 数据包后，根据 IP 数据包上的目的 IP 地址查询路由表，如图 3.14 所示。

MS0 发现路由表中没有关于目的网络（115.0.0.0/29）的路由条目，按规则只能丢弃！解决办法之一是按本项目任务 1 的思路，往路由表中添加静态路由。

```
MS0#sh ip route
Codes: C - connected, S - static, I - IGRP, R - RIP, M - mobile, B - BGP
       D - EIGRP, EX - EIGRP external, O - OSPF, IA - OSPF inter area
       N1 - OSPF NSSA external type 1, N2 - OSPF NSSA external type 2
       E1 - OSPF external type 1, E2 - OSPF external type 2, E - EGP
       i - IS-IS, L1 - IS-IS level-1, L2 - IS-IS level-2, ia - IS-IS inter area
       * - candidate default, U - per-user static route, o - ODR
       P - periodic downloaded static route

Gateway of last resort is not set

     192.168.0.0/30 is subnetted, 1 subnets
C        192.168.0.0 is directly connected, FastEthernet0/24
C    192.168.10.0/24 is directly connected, Vlan10
C    192.168.20.0/24 is directly connected, Vlan20
```

图 3.14 MS0 路由表

2. 在 MS0 路由表中添加静态路由，将发往 ISP 路由器的数据转发给边界路由器 R0

在 MS0 路由表中添加 115.0.0.0/29 的路由条目，命令如下：

`MS0(config)#ip route 115.0.0.0 255.255.255.248 192.168.0.2`

检查路由表，确认已经添加成功，如图 3.15 所示。

```
MS0#sh ip rou
Codes: C - connected, S - static, I - IGRP, R - RIP, M - mobile, B - BGP
       D - EIGRP, EX - EIGRP external, O - OSPF, IA - OSPF inter area
       N1 - OSPF NSSA external type 1, N2 - OSPF NSSA external type 2
       E1 - OSPF external type 1, E2 - OSPF external type 2, E - EGP
       i - IS-IS, L1 - IS-IS level-1, L2 - IS-IS level-2, ia - IS-IS inter area
       * - candidate default, U - per-user static route, o - ODR
       P - periodic downloaded static route

Gateway of last resort is not set

     115.0.0.0/29 is subnetted, 1 subnets
S        115.0.0.0 [1/0] via 192.168.0.2
     192.168.0.0/30 is subnetted, 1 subnets
C        192.168.0.0 is directly connected, FastEthernet0/24
C    192.168.10.0/24 is directly connected, Vlan10
C    192.168.20.0/24 is directly connected, Vlan20
```

图 3.15 核心交换机 MS0 的路由表

MS0 根据新添加的路由条目，将数据转发给 192.168.0.2（边界路由器 R0）。R0 同样也要去查路由表，如图 3.16 所示。

```
R0#sh ip rou
Codes: C - connected, S - static, I - IGRP, R - RIP, M - mobile, B - BGP
       D - EIGRP, EX - EIGRP external, O - OSPF, IA - OSPF inter area
       N1 - OSPF NSSA external type 1, N2 - OSPF NSSA external type 2
       E1 - OSPF external type 1, E2 - OSPF external type 2, E - EGP
       i - IS-IS, L1 - IS-IS level-1, L2 - IS-IS level-2, ia - IS-IS inter area
       * - candidate default, U - per-user static route, o - ODR
       P - periodic downloaded static route

Gateway of last resort is not set

     115.0.0.0/29 is subnetted, 1 subnets
C        115.0.0.0 is directly connected, FastEthernet1/0
     192.168.0.0/30 is subnetted, 1 subnets
C        192.168.0.0 is directly connected, FastEthernet0/0
S    192.168.10.0/24 [1/0] via 192.168.0.1
S    192.168.20.0/24 [1/0] via 192.168.0.1
```

图 3.16 边界路由器 R0 的路由表

路由表会告诉 R0，发往 115.0.0.0/29 的网络，通过 FastEthernet1/0 端口发出。最终数据就到达对端——ISP 的路由器。

显然，通过在内网添加路由条目的方式，可以顺利将数据发送到外网的目的网络。但位于外网的目的网络主机（ISP 路由器）是否能顺利将应答数据包返回来给发送方 Laptop3 呢？我们继续进行应答包返回过程的分析。

3. 分析 ISP 路由器应答包返回的过程

ISP 的路由器接收到数据包以后，会生成一个应答包发回给 Laptop3，地址信息如表 3.5 所示。

表 3.5　ISP 路由器发给 Laptop3 的数据包地址信息

源 IP 地址	目的 IP 地址	目的网络的网络地址
115.0.0.2/29	192.168.20.13/24	192.168.20.0/24

为了把这个数据包发回给 Laptop3，ISP 路由器会去查自己的路由表，以便确定该从哪个端口发出该数据包。ISP 路由器的路由表如图 3.17 所示。

```
ISP#sh ip rou
Codes: C - connected, S - static, I - IGRP, R - RIP, M - mobile, B - BGP
       D - EIGRP, EX - EIGRP external, O - OSPF, IA - OSPF inter area
       N1 - OSPF NSSA external type 1, N2 - OSPF NSSA external type 2
       E1 - OSPF external type 1, E2 - OSPF external type 2, E - EGP
       i - IS-IS, L1 - IS-IS level-1, L2 - IS-IS level-2, ia - IS-IS inter area
       * - candidate default, U - per-user static route, o - ODR
       P - periodic downloaded static route

Gateway of last resort is not set

     115.0.0.0/29 is subnetted, 1 subnets
C       115.0.0.0 is directly connected, FastEthernet1/0
```

图 3.17　ISP 路由器的路由表

但是，从输出的路由信息来看，ISP 路由器的路由表并没有关于 192.168.20.0/24 网络的路由条目。按规则，ISP 会将应答包丢弃，导致数据通信无法继续。

是否可以在 ISP 的路由器添加一条静态路由，让路由器将数据包发回来给 Laptop3 呢？许多人在掌握了静态路由技术以后，很容易会想到要按照本项目任务 1 的思路，在 ISP 路由器添加一条静态路由，让 ISP 将数据发回给 R0 就可以了。

但是这样做会产生两个问题：

①私有 IP 地址在互联网是不可以被路由的，因为公网设备的路由表不允许存在私有 IP 地址的记录。

②私有网络的地址是组织内部自己规划，可以任意使用的。运营商的多个客户如果用的是相同的私有网络地址且允许私有网络地址出现在公网的路由器，那么运营商必须配置到达用户网络的路由条目，而对于某个被重复使用的网络，下一跳地址是不唯一的，最终会导致数据丢失。比如 A 公司和 B 公司都接入了同一个运营商，而且这两个公司都使用同一个私有 IP 地址段（192.168.0.0/24），那么运营商在写路由条目的时候，对于到达 192.168.0.0/24 网络的下一跳节点应该指向 A 公司还是 B 公司呢？如果指向 A 公司，则 B

公司就收不到返回的流量；如果指向 B 公司，也同样导致 A 公司收不到返回的流量；如果写两条，一条指向 A 公司，一条指向 B 公司，则 A 和 B 这两个公司将只会收到 50%的返回流量。显然都不行。

解决问题的办法是应用网络地址转换（Network Address Translation，NAT）技术。边界路由器负责将 Laptop3 发过来的数据包上的源 IP 地址（私有地址）转换成公有 IP 地址。因为 ISP 路由器生成应答包的时候，目的地址是根据它接收到的数据包的源地址确定的（类似于我们收到一封信，回信的时候，目的地址写的是收到信件的源地址）。就这样，当 ISP 的路由器返回应答包的时候，目的地址也是公有地址，避免了私有地址在公网上的非法存在。

端口地址转换（Port Address Translation，PAT）是网络地址转换（Network Address Translation，NAT）技术最常用的一种方式。下面在边界路由器应用 PAT 技术将数据包的源地址进行转换（私有 IP 地址转换成公有 IP 地址）。

步骤 1：定义 NAT 内网端口和外网端口。

对照项目拓扑，边界路由器 R0 的 Fa0/0 端口连接的是企业内网，如图 3.18 所示。

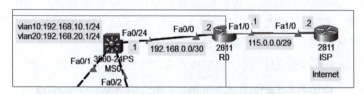

图 3.18　项目拓扑

所以在这里需要将这个端口定义为 NAT 的内网端口，命令如下：

```
R0>
R0>en
R0#conf t
R0(config)#int fa 0/0
R0(config-if)#ip nat inside    //将 Fa0/0 端口定义为 NAT 内网端口
R0(config-if)#exit
R0(config)#
```

路由器 R0 的 Fa1/0 端口连接的是企业外网，所以在这里需要将这个端口定义为 NAT 的外网端口，命令如下：

```
R0(config)#int fa 1/0
R0(config-if)#ip nat outside    //将 Fa1/0 端口定义为 NAT 外网端口
R0(config-if)#exit
```

步骤 2：用访问控制列表（Access Control Lists，ACL）捕获需要转换的流量。这里捕获 Laptop3 所在网络发出的数据包，命令如下：

```
R0(config)#access-list 1 permit 192.168.20.0 0.0.0.255
```

这里创建了一个列表，编号是"1"，这个列表可以捕获网络 192.168.20.0/24 发出的数据包。(编号可以根据情况自己定义,但是要遵守规则。)

注意:这里用到了通配符掩码而不是以前习惯使用的掩码的格式。一种通配符掩码的计算方法,就是简单地将 255.255.255.255 减掉子网掩码。在这个例子中,255.255.255.255 减去 255.255.255.0,就得到 0.0.0.255。

步骤 3:将捕获的流量映射到外网端口,命令如下:

```
R0(config)#ip nat inside source list 1 interface fastEthernet 1/0 overload
```

参数解读:

① "ip nat" 指令是要求路由器执行地址转换。

② "inside source" 参数表示当数据包由内网发往外网的时候,需要转换其源地址。

③ "list 1" 表示编号为 "1" 的访问控制列表。这里要注意的是:列表的编号要和步骤 2 中定义的列表编号一致。

④ "fastEthernet 1/0" 指的是边界路由器 R0 的外网端口,连接运营商的网络端口。

⑤ "overload" 表示允许过载,边界路由器会通过端口映射的方式,允许内网多个用户同时实现地址转换,理论上的数量可以达到 64 511 个 (65 535-1 024)。

步骤 4:测试从内网主机访问外网的情况。

从 Laptop3 ping 外网 ISP 路由器 (115.0.0.2/29),结果如图 3.19 所示。

```
C:\>ping 115.0.0.2

Pinging 115.0.0.2 with 32 bytes of data:

Reply from 115.0.0.2: bytes=32 time<1ms TTL=253
Reply from 115.0.0.2: bytes=32 time=3ms TTL=253
Reply from 115.0.0.2: bytes=32 time=11ms TTL=253
Reply from 115.0.0.2: bytes=32 time=13ms TTL=253
```

图 3.19 Laptop3 ping 外网 ISP 路由器 (115.0.0.2/29) 结果

结果显示通信是正常的,我们通过应用 PAT 技术成功实现了内网用户访问外网。在边界路由器 R0 我们可以查看到地址转换的详细信息,如图 3.20 所示。

```
R0#sh ip nat translations
Pro  Inside global     Inside local       Outside local      Outside global
icmp 115.0.0.1:17      192.168.20.13:17   115.0.0.2:17       115.0.0.2:17
icmp 115.0.0.1:18      192.168.20.13:18   115.0.0.2:18       115.0.0.2:18
icmp 115.0.0.1:19      192.168.20.13:19   115.0.0.2:19       115.0.0.2:19
icmp 115.0.0.1:20      192.168.20.13:20   115.0.0.2:20       115.0.0.2:20
```

图 3.20 边界路由器的地址转换信息

Show ip nat translations 命令显示每个包转换的信息,转换了 4 个 icmp 数据包的地址,包括:数据包从内网出去的时候,将源地址 (192.168.20.13) 转换为内部全局地址 (Inside global:115.0.0.1);数据包从外网进入内网的时候,将目的地址 (115.0.0.1) 转换为内部本地地址 (192.168.20.13)。往返的数据包进行地址转换的时候,是一个相反的过程。

通过 ACL 也可以查看到捕获的数据包统计信息,一共 8 个数据包被捕获,如图 3.21 所示。

```
R0#
R0#sh ip access-lists 1
Standard IP access list 1
    permit 192.168.20.0 0.0.0.255 (8 match(es))

R0#
```

图 3.21　检查边界路由器 R0 上捕获的数据包

边界路由器访问控制列表捕获的数据包统计数据"8 match（es）"：一共有 8 个数据包被匹配到。我们用 ping 命令不是只发出了 4 个包吗？show ip translations 命令显示的信息也只有 4 条记录而已啊。但是，这个捕获的统计数据包括了出去的 4 个数据包和返回的 4 个数据包。（注意：数据包出去和返回都需要转换，只是数据包出去的时候源地址是私有地址，需要转换成公有地址；数据包回来的时候转换的是目的地址，目的地址是公有地址，要转换成原来的私有地址。）

这个时候，销售部 VLAN 10 的用户还是不能访问外网。因为边界路由器 R0 没有去捕获它们网络发出来的流量，也就没有去转换那些数据包的地址。按本任务开始时候的分析，这导致了数据包到达目的地以后，因为应答包的目的地址是一个私有 IP 地址而无法被路由回来，最终被丢弃。

解决的办法很简单，在边界路由器的访问控制列表中添加一条规则，让路由器捕获 VLAN 10 发出的数据即可。

步骤 5：在边界路由器 R0 的访问控制列表中添加规则，让路由器捕获 VLAN 10 网络发出的数据。命令如下：

```
R0(config)#access-list 1 permit 192.168.10.0 0.0.0.255
```

测试销售部用户 Laptop0 访问外网，结果如图 3.22 所示。

```
C:\>ping 115.0.0.2

Pinging 115.0.0.2 with 32 bytes of data:

Reply from 115.0.0.2: bytes=32 time=1ms TTL=253
Reply from 115.0.0.2: bytes=32 time=14ms TTL=253
Reply from 115.0.0.2: bytes=32 time=15ms TTL=253
Reply from 115.0.0.2: bytes=32 time=12ms TTL=253
```

图 3.22　Laptop0 访问 ISP 路由器应答信息

结果显示，销售部用户的计算机也能正常访问外网。

课堂练习：参照本任务，在边界路由器配置 PAT，让企业内网所有用户都能访问到 ISP 的路由器。注：ISP 路由器上不能有任何内网的路由。

素质拓展：运营商为用户提供互联网的接入服务。人们可以通过网络获取信息的方式更加方便、多样，大部分人特别是年轻人越来越主要地依靠网络获取信息。与此同时，网上也充斥着越来越多的负面信息内容，影响了网络生活秩序。大学生应当正确使用网络，提高信息的获取能力，加强信息的辨识能力，增进信息的应用能力，使网络成为开阔视野、提高能力的重要工具。

任务 3：实现互联网用户访问内网主机——网络地址转换（静态 NAT）

实现互联网用户访问内网主机——网络地址转换（静态 NAT）

任务目标：应用静态 NAT 技术，实现外网用户能访问内网主机。

上一个任务我们用 PAT 技术实现了内网用户访问外网主机，如果外网用户想访问内网主机，如何实现呢？有一种技术可以实现这个需求，那就是静态 NAT 技术，即通过私有 IP 地址和公有 IP 地址一对一映射的方式实现。

下面，我们通过配置静态 NAT 技术，实现 ISP 的路由器能 ping 通 Laptop3。

步骤 1：定义 NAT 内网端口和外网端口。

依据项目拓扑，边界路由器 R0 的 Fa0/0 端口连接的是企业内网，如图 3.23 所示。

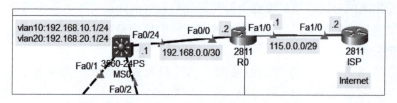

图 3.23　项目拓扑

所以在这里需要将 R0 的 Fa0/0 端口定义为 NAT 的内网端口，命令如下：

```
R0>
R0>enable
R0#conf t
R0(config)#int fa0/0
R0(config-if)#ip nat inside      //将 Fa0/0 端口定义为 NAT 内网端口
R0(config-if)#exit
R0(config)#
```

路由器 R0 的 Fa1/0 端口连接的是企业内网，所以在这里需要将该端口定义为 NAT 的外网端口，命令如下：

```
R0(config)#int fa1/0
R0(config-if)#ip nat outside     //将 Fa1/0 端口定义为 NAT 外网端口
R0(config-if)#exit
```

注：如果在本项目任务 2 中已经定义好 NAT 内网端口和外网端口，则不用重复配置，可以跳过此步骤。

步骤 2：将私有 IP 地址映射到公有 IP 地址。

在全局模式下，将内网 Laptop3 的 IP 地址（192.168.20.13）映射到外网端口所在网络（115.0.0.0/29）的某个地址。

经过计算，网络 115.0.0.0/29 的主机地址范围为 115.0.0.1/29～115.0.0.6/29。由于 115.0.0.1 和 115.0.0.2 已经被 R0 和 ISP 这两台设备占用，这里用该网络目前还未被使用的主机地址：115.0.0.3。

配置静态 NAT 映射的命令如下：

```
R0(config)#ip nat inside source static 192.168.20.13 115.0.0.3
```

命令解读：
① "static"参数是静态的意思。
② "192.168.20.13"是内网主机 Laptop3 的 IP 地址。
③ "115.0.0.3"是边界路由器外网端口同网段的合法公有地址。（可以使用外网端口的 IP 地址。）

这样，外部网络的用户就可以访问到公有地址 115.0.0.3，然后边界路由器将外网发来的数据转发给内网的 192.168.20.13 主机，从而实现外网用户能访问到内网主机资源。

步骤 3：测试 ISP 路由器访问 Laptop3 的情况。

注意，ISP 路由器直接 ping Laptop3 的 IP 地址是不成功的，因为外网一定没有私有地址的路由条目。而此时，内网 Laptop3 的地址（192.168.20.13）已经映射为外网可以识别的地址（115.0.0.3），所以测试的时候，应该从外网 ping 映射以后的外网地址（115.0.0.3），如图 3.24 所示。

```
ISP#ping 115.0.0.3

Type escape sequence to abort.
Sending 5, 100-byte ICMP Echos to 115.0.0.3, timeout is 2 seconds:
!!!!!
Success rate is 100 percent (5/5), round-trip min/avg/max = 0/13/28 ms
```

图 3.24　ISP 路由器 ping Laptop3 主机映射的地址结果

结果显示通信是正常的。在边界路由器中查看转换记录，结果如图 3.25 所示。

```
R0#show ip nat translations
Pro   Inside global    Inside local     Outside local     Outside global
---   115.0.0.3        192.168.20.13    ---               ---

R0#
```

图 3.25　边界路由器的 NAT 映射表

我们开启边界路由器 R0 的 debug 功能，如图 3.26 所示。

```
R0#debug ip nat
IP NAT debugging is on
R0#
```

图 3.26　开启 R0 的 debug 功能

ISP 路由器再次 ping 主机 Laptop3 映射的地址，然后查看路由器 R0 的 debug 信息，如图 3.27 所示。

```
R0#debug ip nat
IP NAT debugging is on
R0#
NAT: s=115.0.0.2, d=115.0.0.3->192.168.20.13 [47]

NAT*: s=192.168.20.13->115.0.0.3, d=115.0.0.2 [40]

NAT: s=115.0.0.2, d=115.0.0.3->192.168.20.13 [48]

NAT*: s=192.168.20.13->115.0.0.3, d=115.0.0.2 [41]

NAT: s=115.0.0.2, d=115.0.0.3->192.168.20.13 [49]

NAT*: s=192.168.20.13->115.0.0.3, d=115.0.0.2 [42]

NAT: s=115.0.0.2, d=115.0.0.3->192.168.20.13 [50]

NAT*: s=192.168.20.13->115.0.0.3, d=115.0.0.2 [43]

NAT: s=115.0.0.2, d=115.0.0.3->192.168.20.13 [51]

NAT*: s=192.168.20.13->115.0.0.3, d=115.0.0.2 [44]
```

图 3.27　R0 路由器 debug 信息

信息解读：

① "NAT"表示数据包从外网进入内网时的转换信息。

② "NAT＊"表示数据包从内网发出外网时的转换信息。

③ "d = 115.0.0.3->192.168.20.13"的 d 表示目的（Destination）地址，指的是当数据包从外网进入内网的时候，将数据包的目的地址 115.0.0.3 转换为 192.168.20.13。

④ "s = 192.168.20.13->115.0.0.3"的 s 表示源（Source）地址，指的是当应答包从内网返回外网的时候，将源地址 192.168.20.13 转换为 115.0.0.3。

⑤ 最后的中括号［47］表示 NAT 转换时使用的端口号为 47。

静态的地址映射可以精确到端口号。这个案例中，我们将两个 IP 地址相互映射，这两个 IP 地址所有的端口号也都是一一对应的。当然，如果只是想映射某个特定端口号，也是可以的。下面的命令可以将内网地址（192.168.20.13）的 8080 端口映射到外网地址（115.0.0.3）的 80 端口，命令如下：

```
R0(config)#ip nat inside source static tcp 192.168.20.13 8080 115.0.0.3 80
```

课堂练习：参照本任务，在企业内网增加一台服务器，接入核心交换机，适当配置，让服务器能访问到边界路由器。然后在边界路由器 R0 中配置 PAT，让 Internet 用户能访问到企业内网的服务器。

素质拓展：静态 NAT 映射技术，通常应用于为实现部署于内部网络的服务器能对外网用户提供服务的情景。但是该技术让内部服务器暴露于 Internet 而可能受到很大的外部威胁。工程人员需要严格遵循 GB/T 22239—2019《信息安全技术 网络安全等级保护基本要求》标准，为信息系统构筑坚固的堡垒，防止信息泄露。大学生要筑牢自己思想的堡垒，要增强安全意识，对境内外敌对势力的渗透、颠覆、破坏活动保持高度警惕，切实履行维护国家安全的义务。

任务 4：减少路由表的路由条目——路由汇总（默认路由）

任务目标：利用一条默认路由，将所有内网的数据发到外网。

在本项目任务 1 中，为了让内网的主机能访问外网中某个网络的主机，就需要在内网的所有三层设备（核心交换机和边界路由器）中添加路由条目，否则三层设备在自己的路由表没有查到目的网络的相关路由信息，就会把数据丢弃。

减少路由表的路由条目——路由汇总（默认路由）

为了能够更好地理解这个知识点，我们把项目拓扑的 ISP 部分补齐：增加来自运营商互联网数据中心（Internet Data Center，IDC）的两台服务器，如图 3.28 所示。

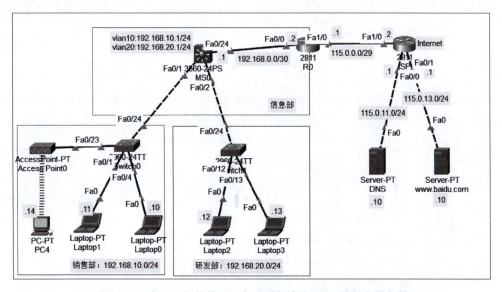

图 3.28　在项目拓扑的 ISP 部分增加来自 IDC 的两台服务器

1. 配置服务器的 IP 地址以及网关，让外网设备能相互通信

IDC 各设备的 IP 地址信息如表 3.6 所示。

表 3.6　IDC 各设备的 IP 地址信息

设备名称	端口	IP 地址	备注
DNS 服务器		115.0.11.10/24	
Web 服务器		115.0.13.10/24	
ISP 路由器	Fa0/0	115.0.11.1/24	DNS 服务器的网关
	Fa0/1	115.0.13.1/24	Web 服务器的网关

2. 在内网的三层设备中添加路由，将内网的数据发往 ISP 数据中心服务器

为了让内网用户能访问到外网的两个网络 115.0.11.0/24 和 115.0.13.0/24，我们必须

在内网的核心交换机中添加两条路由，让核心交换机能将发往这两个网络的数据转发给 R0，MS0 的路由表如图 3.29 所示。

```
MS0#sh ip rou
Codes: C - connected, S - static, I - IGRP, R - RIP, M - mobile, B - BGP
       D - EIGRP, EX - EIGRP external, O - OSPF, IA - OSPF inter area
       N1 - OSPF NSSA external type 1, N2 - OSPF NSSA external type 2
       E1 - OSPF external type 1, E2 - OSPF external type 2, E - EGP
       i - IS-IS, L1 - IS-IS level-1, L2 - IS-IS level-2, ia - IS-IS inter area
       * - candidate default, U - per-user static route, o - ODR
       P - periodic downloaded static route

Gateway of last resort is not set

     115.0.0.0/8 is variably subnetted, 3 subnets, 2 masks
S       115.0.0.0/29 [1/0] via 192.168.0.2
S       115.0.11.0/24 [1/0] via 192.168.0.2
S       115.0.13.0/24 [1/0] via 192.168.0.2
     192.168.0.0/30 is subnetted, 1 subnets
C       192.168.0.0 is directly connected, FastEthernet0/24
C    192.168.10.0/24 is directly connected, Vlan10
C    192.168.20.0/24 is directly connected, Vlan20
```

图 3.29　MS0 的路由表

同时也要在边界路由器 R0 中添加两条路由，让 R0 将发往 115.0.11.0/24 和 115.0.13.0/24 网络的数据转发给 ISP 的路由器，R0 的路由表如图 3.30 所示。

```
R0(config)#do sh ip rou
Codes: C - connected, S - static, I - IGRP, R - RIP, M - mobile, B - BGP
       D - EIGRP, EX - EIGRP external, O - OSPF, IA - OSPF inter area
       N1 - OSPF NSSA external type 1, N2 - OSPF NSSA external type 2
       E1 - OSPF external type 1, E2 - OSPF external type 2, E - EGP
       i - IS-IS, L1 - IS-IS level-1, L2 - IS-IS level-2, ia - IS-IS inter area
       * - candidate default, U - per-user static route, o - ODR
       P - periodic downloaded static route

Gateway of last resort is not set

     115.0.0.0/8 is variably subnetted, 3 subnets, 2 masks
C       115.0.0.0/29 is directly connected, FastEthernet1/0
S       115.0.11.0/24 [1/0] via 115.0.0.2
S       115.0.13.0/24 [1/0] via 115.0.0.2
     192.168.0.0/30 is subnetted, 1 subnets
C       192.168.0.0 is directly connected, FastEthernet0/0
S    192.168.10.0/24 [1/0] via 192.168.0.1
S    192.168.20.0/24 [1/0] via 192.168.0.1
```

图 3.30　R0 的路由表

检查从内网访问外网服务器。我们从 Laptop3 主机 ping 外网的 Web 服务器（115.0.13.10/24），结果显示通信正常，如图 3.31 所示。

```
C:\>ping 115.0.13.10

Pinging 115.0.13.10 with 32 bytes of data:

Reply from 115.0.13.10: bytes=32 time<1ms TTL=125
Reply from 115.0.13.10: bytes=32 time<1ms TTL=125
Reply from 115.0.13.10: bytes=32 time=11ms TTL=125
Reply from 115.0.13.10: bytes=32 time=15ms TTL=125
```

图 3.31　Laptop3 主机 ping 外网的 Web 服务器结果

问题是，整个互联网有数百万个网络，难道要在内网的每台三层设备中都添加所有的几百万条路由信息吗？如果真的要这样做，那会导致几乎所有内网的三层设备无法承受如此庞大的路由表。对于聪明的人类而言，该解决方案显然是无法被接受的。

路由汇总可以解决这个问题。我们尝试将这两条路由的网络汇总成更大的网络，在路由表中用汇总后的网络来表达路由信息，可以减少路由表的条目，提高查询效率。

3. 路由汇总：将多个网络的路由条目汇总成一条

步骤1：将需要汇总的路由条目的目标网络转换成二进制表达方式，如图3.32所示。

目标网络号	二进制表达	备注
115.0.11.0	01110011.00000000.00001 011.00000000	
115.0.13.0	01110011.00000000.00001 101.00000000	

网络位部分　　　　　主机位部分

图 3.32　转换成二进制表达方式

步骤2：将网络位部分和主机位部分分开，得到汇总后的网络。

对照这两个网络地址的二进制数表达形式，将相同的部分和不同的部分隔开。相同的部分是汇总后的网络的网络位部分，不同的部分是汇总后的网络的主机位部分。将主机位部分全部写成"0"，就得到汇总后的网络的网络号，如表3.7所示。

表 3.7　汇总后的网络的网络号

网络位部分	主机位部分	十进制表达
01110011.00000000.00001	000.00000000	115.0.8.0/21

由此可知网络115.0.11.0/24和115.0.13.0/24汇总后，得到一个更大的网络，网络号为115.0.8.0/21。也就是说，在路由表中，我们用以下命令，用汇总以后的路由，替换掉原来的两条路由记录。

步骤3：在三层设备中用汇总后的路由替换原来的明细路由。

先删除MS0和R0上网络115.0.11.0/24和115.0.13.0/24的明细路由，命令如下：

```
MS0(config)#no ip route 115.0.11.0 255.255.255.0 192.168.0.2
MS0(config)#no ip route 115.0.13.0 255.255.255.0 192.168.0.2

R0(config)#no ip route 115.0.11.0 255.255.255.0 115.0.0.2
R0(config)#no ip route 115.0.13.0 255.255.255.0 115.0.0.2
```

在MS0和R0上配置汇总后的路由条目，命令如下：

```
MS0(config)#ip route 115.0.8.0 255.255.248.0 192.168.0.2
```

它表达的是：发往115.0.8.0/21网络的数据，一律转发给192.168.0.2。

```
R0(config)#ip route 115.0.8.0 255.255.248.0 115.0.0.2
```

它表达的是：发往115.0.8.0/21网络的数据，一律转发给115.0.0.2。

由于网络 115.0.11.0/24 和 115.0.13.0/24 是包含在网络 115.0.8.0/21 里的，所以汇总后的路由同样也可以转发目的网络为 115.0.11.0/24 和 115.0.13.0/24 的数据。

核心交换机 MS0 汇总后的路由表如图 3.33 所示。

```
MS0#sh ip rou
Codes: C - connected, S - static, I - IGRP, R - RIP, M - mobile, B - BGP
       D - EIGRP, EX - EIGRP external, O - OSPF, IA - OSPF inter area
       N1 - OSPF NSSA external type 1, N2 - OSPF NSSA external type 2
       E1 - OSPF external type 1, E2 - OSPF external type 2, E - EGP
       i - IS-IS, L1 - IS-IS level-1, L2 - IS-IS level-2, ia - IS-IS inter area
       * - candidate default, U - per-user static route, o - ODR
       P - periodic downloaded static route

Gateway of last resort is not set

     115.0.0.0/8 is variably subnetted, 2 subnets, 2 masks
S       115.0.0.0/29 [1/0] via 192.168.0.2
S       115.0.8.0/21 [1/0] via 192.168.0.2
     192.168.0.0/30 is subnetted, 1 subnets
C       192.168.0.0 is directly connected, FastEthernet0/24
C    192.168.10.0/24 is directly connected, Vlan10
C    192.168.20.0/24 is directly connected, Vlan20
```

图 3.33　核心交换机 MS0 汇总后的路由表

边界路由器 R0 汇总后的路由表如图 3.34 所示。

```
R0#sh ip rou
Codes: C - connected, S - static, I - IGRP, R - RIP, M - mobile, B - BGP
       D - EIGRP, EX - EIGRP external, O - OSPF, IA - OSPF inter area
       N1 - OSPF NSSA external type 1, N2 - OSPF NSSA external type 2
       E1 - OSPF external type 1, E2 - OSPF external type 2, E - EGP
       i - IS-IS, L1 - IS-IS level-1, L2 - IS-IS level-2, ia - IS-IS inter area
       * - candidate default, U - per-user static route, o - ODR
       P - periodic downloaded static route

Gateway of last resort is not set

     115.0.0.0/8 is variably subnetted, 2 subnets, 2 masks
C       115.0.0.0/29 is directly connected, FastEthernet1/0
S       115.0.8.0/21 [1/0] via 115.0.0.2
     192.168.0.0/30 is subnetted, 1 subnets
C       192.168.0.0 is directly connected, FastEthernet0/0
S    192.168.10.0/24 [1/0] via 192.168.0.1
S    192.168.20.0/24 [1/0] via 192.168.0.1
```

图 3.34　边界路由器 R0 汇总后的路由表

这里我们只是将两条路由汇总成一条。可是互联网有好几百万条路由，该怎样汇总呢？

4. 再汇总：将所有网络的路由条目汇总成一条

我们可以从这个汇总的小案例中寻找规律，然后做出推断。如表 3.8 所示是 Internet 区域汇总前后 IP 地址信息表。

表 3.8　Internet 区域汇总前后 IP 地址信息表

汇总前		汇总后	
网络号	子网掩码	网络号	子网掩码
115.0.11.0/24	/24	115.0.8.0	/21
115.0.13.0/24	/24		

从表 3.8 可以看出，汇总前的子网掩码长度为/24，汇总后子网掩码的长度为/21——子网掩码的长度变短了。

按这样的规律去推断，网络越大其子网掩码就应该越短，当网络大到极限的时候，子网掩码也应该短到极限。因此，全球所有组织的网络汇总后，得到全球最大的网络（Internet），其子网掩码必然为最短（/0）。

子网掩码为/0，表示这个网络没有网络位，全部都是主机位。而网络地址的主机位全是"0"，所以要表达 Internet 的所有汇总后的网络，可以用 0.0.0.0/0 表示。也就是说，我们可以写一条路由，将全部数据发往 Internet。

首先删除 MS0 和 R0 上的汇总静态路由。

在 MS0 和 R0 上配置汇总后的路由条目，命令如下：

```
MS0(config)#no ip route 115.0.8.0 255.255.248.0 192.168.0.2
R0(config)#no ip route 115.0.8.0 255.255.248.0 115.0.0.2
```

在核心交换机 MS0 中配置新汇总的路由，命令如下：

```
MS0(config)#ip route 0.0.0.0 0.0.0.0 192.168.0.2    //将所有的数据都转发给 192.168.0.2 节点
```

在边界路由器 R0 中配置新汇总的路由，命令如下：

```
R0(config)#ip route 0.0.0.0 0.0.0.0 115.0.0.2    //将所有的数据都转发给 115.0.0.2 节点
```

替换后核心交换机 MS0 上的路由表如图 3.35 所示。

```
MS0#sh ip rou
Codes: C - connected, S - static, I - IGRP, R - RIP, M - mobile, B - BGP
       D - EIGRP, EX - EIGRP external, O - OSPF, IA - OSPF inter area
       N1 - OSPF NSSA external type 1, N2 - OSPF NSSA external type 2
       E1 - OSPF external type 1, E2 - OSPF external type 2, E - EGP
       i - IS-IS, L1 - IS-IS level-1, L2 - IS-IS level-2, ia - IS-IS inter area
       * - candidate default, U - per-user static route, o - ODR
       P - periodic downloaded static route

Gateway of last resort is 192.168.0.2 to network 0.0.0.0

     115.0.0.0/29 is subnetted, 1 subnets
S       115.0.0.0 [1/0] via 192.168.0.2
     192.168.0.0/30 is subnetted, 1 subnets
C       192.168.0.0 is directly connected, FastEthernet0/24
C       192.168.10.0/24 is directly connected, Vlan10
C       192.168.20.0/24 is directly connected, Vlan20
S*   0.0.0.0/0 [1/0] via 192.168.0.2
```

图 3.35　替换后核心交换机 MS0 上的路由表

边界路由器 R0 上的路由表如图 3.36 所示。

路由表中标记 0.0.0.0/0 的静态路由不再标记为"S"，而是标记为"S*"，称为静态默认路由。

这里用静态默认路由将内网发到 Internet 的所有数据往运营商发送，极大地精简了内网

```
R0#sh ip rou
Codes: C - connected, S - static, I - IGRP, R - RIP, M - mobile, B - BGP
       D - EIGRP, EX - EIGRP external, O - OSPF, IA - OSPF inter area
       N1 - OSPF NSSA external type 1, N2 - OSPF NSSA external type 2
       E1 - OSPF external type 1, E2 - OSPF external type 2, E - EGP
       i - IS-IS, L1 - IS-IS level-1, L2 - IS-IS level-2, ia - IS-IS inter area
       * - candidate default, U - per-user static route, o - ODR
       P - periodic downloaded static route

Gateway of last resort is 115.0.0.2 to network 0.0.0.0

     115.0.0.0/29 is subnetted, 1 subnets
C       115.0.0.0 is directly connected, FastEthernet1/0
     192.168.0.0/30 is subnetted, 1 subnets
C       192.168.0.0 is directly connected, FastEthernet0/0
S    192.168.10.0/24 [1/0] via 192.168.0.1
S    192.168.20.0/24 [1/0] via 192.168.0.1
S*   0.0.0.0/0 [1/0] via 115.0.0.2
```

图 3.36 边界路由器 R0 上的路由表

三层设备的路由表，极大地提高了查询路由表的效率，同时也降低了网络的管理成本。还有一个好处，那就是如果 Internet 多出了其他的网络，不需要网络管理员去了解多出了哪些网络，其网络号是多少，更不需要在内网的三层设备额外添加任何路由条目，因为一条默认路由就可以将所有数据往外网发送。

但需要注意的是，默认路由的优先级低于明细路由，因为路由器查询路由表是根据子网掩码的最长匹配原则进行的，而默认路由的子网掩码为 0，按逻辑，当然是最后才会匹配到。也就是说，路由器根据数据包上的目的地址去路由表匹配路由条目，如果明细路由条目匹配不到，最后才匹配到默认路由，而此时的默认路由因为可以被所有目的网络匹配到，所以路由器会按默认路由的下一跳 IP 地址转发数据，而不会导致数据包在这台路由器上丢失。

课堂练习：参照本任务，将内网中目的网络为公有地址的静态路由全部删除，用默认路由替代，配置完成后，内网用户依然可以访问到 ISP 的路由器。

素质拓展：路由汇总简化了路由表的路由条目，提高了路由器匹配路由条目的效率，从而提高了数据转发速率。简单是一种美，大学生要身体力行，倡导简约适度、绿色低碳的生活方式，像对待生命一样对待生态环境，为留下天蓝、地绿、水清的生产生活环境，为建设美丽中国作出自己应有的贡献。

任务 5：部署域名服务器——域名服务（DNS）

部署域名服务器——域名服务(DNS)

任务目标：理解通过域名访问一个 Web 服务器的过程。

完成本项目任务 1 到 3 的配置后，我们可以在 Laptop3 用浏览器在 URL 地址栏输入 Web 服务器的 IP 地址，访问外网的 Web 服务器，结果如图 3.37 所示。

我们如何确认这里访问到的网页就是 Web 服务器（115.0.13.10/24）上的呢？我们可以在 Web 服务器上修改这个网页，如图 3.38、图 3.39 所示。

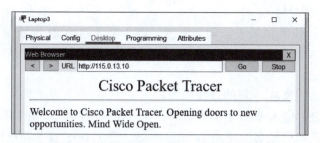

图 3.37　Laptop3 访问外网的 Web 服务器

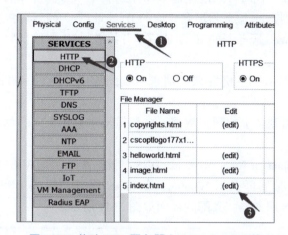

图 3.38　修改 Web 服务器上 index.html 文件

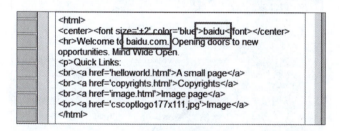

图 3.39　修改默认首页信息

将原网页的两处"Cisco Packet Tracer"分别改成"baidu"和"baidu.com",单击右下角"Save"保存按钮。其间提示是否覆盖(overwrite),单击"Yes"按钮确认覆盖。再次在 Laptop3 上用浏览器访问 Web 服务器,看到页面也跟着改变,如图 3.40 所示。

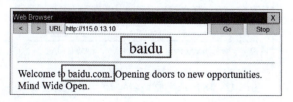

图 3.40　在 Laptop3 上再次访问 Web 服务器

显示的网页随着之前的修改而产生变化,则可以确定在 Laptop3 上看到的网页就是 Web

服务器（115.0.13.10/24）上的。

一般情况下不会用IP地址去访问服务器，因为服务器的IP地址没有规律，不容易记忆。所以我们通过Web主机的名字（域名）来访问。就像同学之间一般称呼姓名，而非学号，因为学号不好记。

用户通过域名访问Web主机的过程如下：

①用户通过地址栏输入Web服务器域名，主机获取用户要访问的域名。

②用户主机将获取的域名发送给DNS服务器，请求DNS服务器查询该域名对应的IP地址，因为主机封装数据包时目的地址是Web服务器的IP地址，而不是其域名。

③DNS服务器查到Web服务器域名对应的IP地址后，会将该IP地址发回给用户主机。

④用户主机得到Web服务器的IP地址后，开始进行IP数据包的封装，将Web请求发给Web服务器。

⑤Web服务器根据用户主机发来的请求，找到相应的网页，将网页数据发回给用户主机。

⑥最终用户主机得到请求的网页数据，在浏览器中显示出来，用户就可以看到相应的网页了。

域名服务系统（Domain Name System，DNS）服务器保存有Web主机的域名和其IP地址对应的表。通常情况下可以用Windows或Linux服务器提供DNS服务，这里用Cisco Packet Tracer模拟DNS服务器，其配置界面如图3.41所示。

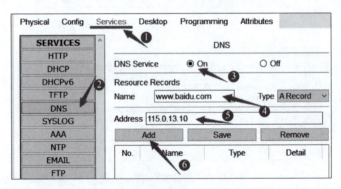

图3.41　DNS服务配置界面

输入域名www.baidu.com和Web服务器的IP地址，单击"Add"按钮往数据库中添加记录，完成后数据库中会增加相应记录，如图3.42所示。

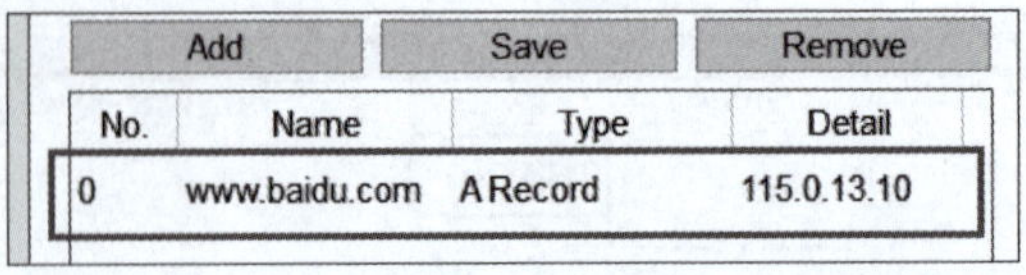

图3.42　DNS服务器域名和IP地址的映射表

添加成功以后，还需要在用户设置网卡参数的位置填写DNS服务器的IP地址，告诉用户主机，如果有域名需要解析成IP地址的，请将域名发送给115.0.11.10/24，如图3.43所示。

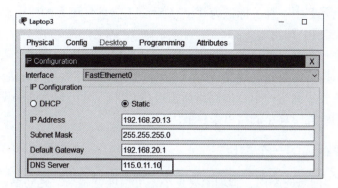

图 3.43　配置 DNS 服务器的 IP 地址

完成所有的配置以后，用户就可以在 Laptop3 上的浏览器地址栏中输入域名访问到 Web 服务器，结果如图 3.44 所示。

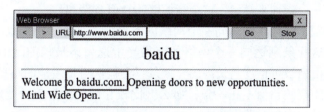

图 3.44　用户通过域名访问 Web 服务器

结果显示，用户可以借助 DNS 服务用域名来访问到 Web 服务器的资源。

课堂练习：参照本任务，在 Internet 部署 Web 和 DNS 服务器，适当配置，实现内网用户能借助 DNS 服务用域名来访问 Web 服务器的资源。

素质拓展：在互联网中我们用 IP 地址和域名来标识一台服务器，但由于域名更方便记忆，所以我们借助 DNS 服务器帮我们将域名解析成 IP 地址，以便 IP 数据包能封装成功。因此，在用户对 DNS 服务如此依赖的情况下，如果 DNS 服务器被劫持，将会导致大量网络服务陷入瘫痪。这种破坏网络通信安全的行为是违法的，大学生要弘扬社会主义法治精神，建设社会主义法治文化，增强全社会厉行法治的积极性和主动性，形成守法光荣、违法可耻的社会氛围。

小结与拓展

1. 静态路由信息解读

```
S 115.0.0.0 [1/0] via 192.168.0.2
```

① "S" 是 "static" 的简写，表示静态路由的意思。
② "115.0.0.0" 表示目标网络的网络号。
③ "[1/0]" 中的 "1" 表示该路由条目的管理距离，"0" 表示该路由条目的度量值。
④ "192.168.0.2" 表示下一跳节点的 IP 地址。

路由表中存放的路由条目都是路由器认为到达目标网络的最佳路径。路由条目有很多种类，静态路由只是其中一种，标记为"S"。后面还会学到更多的路由协议，比如 R 表示 RIP 路由，O 表示 OSPF 路由等。

管理距离代表一个路由条目的可信度。对于到达同一网络的路由，管理距离越小，可信度就越高，表示该路由条目就越优。而只有被认为最优的路由才可以进入路由表中。

路由的度量值表示到达一个目的网络所需要的成本。对于到达同一网络的路由，如果管理距离一样，则管理距离小的路由条目最优。

2. 关于管理距离的一个类比

"对于达到同一网络的路由，管理距离越小，可信度就越高"这句话，有一个比较好的例子帮助我们去理解：有人在陌生的城市迷路了，于是在十字路口问路，去某地方怎么走，连续问了好几个人，他们都是指向不同的方向，此时，迷路的这个人该相信谁？答案是：谁的可信度高就相信谁。

可信度也要有一个指标来衡量，这个指标就叫"管理距离"。

管理员如果配置了好几条到达同一个网络的条目，而且下一跳都是不一样的，这个时候路由器该如何在这几条到达同一个网络的路径当中选一条最佳的路由呢？路由器认为，管理距离最小的路由条目可信度是最高的，该条目会被认为是最佳的路径而被选入路由表；如果当中最小管理距离的路由不止一条，那么就比较它们的度量值，度量值最小的被认为是最佳的路径，会被选中放入路由表中。

3. 最长匹配原则与默认路由在路由表中的地位

路由器匹配路由条目的方式叫"最长匹配原则"。路由器接收到一个数据包后，会读取数据包上的目的地址，去匹配路由表中的网络号。如果查到多个网络号一样的条目，则会匹配到子网掩码最长的那条记录。

按照这个规则，对于子网掩码为 0 的默认路由，总是最后才会被匹配到，因为 0 是最短的子网掩码。

4. 修改静态路由的管理距离

静态路由默认的管理距离是 0，但有需要的时候是可以修改的。以下就是将去往 115.0.11.0/24 网络的静态路由的管理距离改成 100 的命令：

```
MS0(config)# ip route 115.0.11.0 255.255.255.0 192.168.0.2 100
```

检查 MS0 的路由表，结果如图 3.45 所示。

```
Gateway of last resort is 192.168.0.2 to network 0.0.0.0

     115.0.0.0/8 is variably subnetted, 2 subnets, 2 masks
S       115.0.0.0/29 [1/0] via 192.168.0.2
S       115.0.11.0/24 [100/0] via 192.168.0.2
     192.168.0.0/30 is subnetted, 1 subnets
C       192.168.0.0 is directly connected, FastEthernet0/24
C       192.168.10.0/24 is directly connected, Vlan10
C       192.168.20.0/24 is directly connected, Vlan20
S*   0.0.0.0/0 [1/0] via 192.168.0.2
```

图 3.45　MS0 的路由表

从路由表中可以看到，去往 115.0.11.0/24 网络的路由条目的管理距离已经改成了 100。一般在多出口的网络，我们会配置多条管理距离不一样的静态路由。正常情况下启用管理距离小的路由，而当该路由失效，管理距离第二小的路由被启用，从而实现路径的备份，这叫浮动静态路由技术。

5. 用转出端口配置静态路由

静态路由有两种配置方式：带下一跳的静态路由和带转出端口的静态路由。本项目任务 1 中我们采用的是第一种方式，也是最常用的方式，只要下一跳的地址可达，该路由就可用，缺点是需要进行递归查询，占用更多系统资源。

以下是采用带转出端口的静态路由的配置方式。根据如图 3.1 所示的拓扑，MS0 如果有数据要发往 115.0.0.0/29，则数据应该从 Fa0/24 端口发出。于是可以这样配置：

```
MS0(config)#ip route 115.0.0.0 255.255.255.248 fastEthernet 0/24
```

检查核心交换机 MS0 的路由表，结果如图 3.46 所示。

```
MS0(config)#do sh ip rou
Codes: C - connected, S - static, I - IGRP, R - RIP, M - mobile, B - BGP
       D - EIGRP, EX - EIGRP external, O - OSPF, IA - OSPF inter area
       N1 - OSPF NSSA external type 1, N2 - OSPF NSSA external type 2
       E1 - OSPF external type 1, E2 - OSPF external type 2, E - EGP
       i - IS-IS, L1 - IS-IS level-1, L2 - IS-IS level-2, ia - IS-IS inter area
       * - candidate default, U - per-user static route, o - ODR
       P - periodic downloaded static route

Gateway of last resort is 192.168.0.2 to network 0.0.0.0

     115.0.0.0/8 is variably subnetted, 2 subnets, 2 masks
S       115.0.0.0/29 is directly connected, FastEthernet0/24
S       115.0.11.0/24 [100/0] via 192.168.0.2
     192.168.0.0/30 is subnetted, 1 subnets
C       192.168.0.0 is directly connected, FastEthernet0/24
C    192.168.10.0/24 is directly connected, Vlan10
C    192.168.20.0/24 is directly connected, Vlan20
S*   0.0.0.0/0 [1/0] via 192.168.0.2
```

图 3.46 核心交换机 MS0 的路由表

"S 115.0.0.0/29 is directly connected, FastEthernet0/24" 路由条目会告诉交换机 MS0，如果有数据要发往 115.0.0.0/29 网络，则可以将数据从 FastEthernet 0/24 端口转发出去，最终效果是一样的。好处是路由器从该路由条目就直接知道了要将数据从端口 FastEthernet 0/24 转发出去，不用进行递归查询，提升了数据转发效率。

用转出端口配置静态路由有两个缺点。一是当转出端口失效，即使下一跳地址可达，该路径也会失效；二是这里虽然被标记为 S（静态路由），但是后面却显示 115.0.0.0/29 网络是一个直连网络（directly connected），这显然是一个错误。会让交换机 MS0 误以为网络 115.0.0.0/29 跟它是直连的，会发出 ARP 请求，询问目的主机的 MAC 地址。但实际上目的主机并不是直连的，而是需要跨几个网络。

ARP 请求是以广播的方式发出的，如果转发端口是一个广播型的网络，会导致网络上出现过多的流量，可能会耗尽路由器的内存，从而影响正常的网络通信。解决问题的办法是：用 ip route 115.0.0.0 255.255.255.248 fastEthernet0/24 192.168.0.2 命令同时指定转发端口和下一跳地址（Cisco Packet Tracer 模拟器不支持该命令）。

6. NAT 的几种类型

NAT 有三种类型：静态 NAT、动态 NAT 和 PAT。

本项目用到了两种 NAT 技术：静态 NAT 和 PAT。这也是最常用的 NAT 技术。静态 NAT 用于将内部服务器的 IP 地址映射到某个公有 IP 地址，实现外网用户能访问内网服务器；PAT 技术用于实现内部用户主机能访问外网资源，其通过端口映射的方式实现公有 IP 地址复用，节约了公有 IP 地址。

另外一种动态 NAT 的实现思路是：创建一个公有 IP 地址池，将内网用户的 IP 地址映射到地址池中某个空闲的公有地址，命令如下：

```
R0(config)# ip nat pool NAT_POOL 115.0.0.3 115.0.0.6 netmask 255.255.255.248    //定义一个地址池,名字为 NAT_POOL(任意取值)
R0(config)# ip nat inside source list 1 pool NAT_POOL    //将列表定义的用户 IP 映射到地址池(注意列表序号和地址池名称必须和定义的完全匹配)
```

如果地址池中没有空闲的公有 IP 地址，则用户数据只能等待，从而影响到数据通信效率，解决这个问题的办法，只能是注意观察地址池的使用情况，避免枯竭的情况发生，或者用 overload 参数，允许过载转换，命令如下：

```
R0(config)# ip nat inside source list 1 pool NAT_POOL overload    //允许 NAT 过载
```

7. 域名服务

域名服务采用 DNS 协议来实现，其采用分层系统创建数据库以提供域名解析，如图 3.47 所示。DNS 使用域名来划分层次。

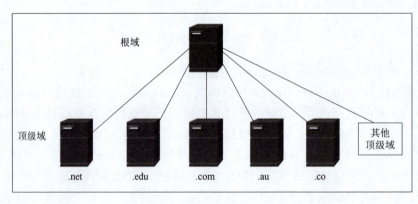

图 3.47　域名结构

域名结构被划分为多个更小的受管域。每台 DNS 服务器维护着特定的数据库文件，而且只负责管理 DNS 结构中那一小部分的"域名—IP"映射。当 DNS 服务器收到的域名转换请求不属于其所负责的 DNS 区域时，该 DNS 服务器可将请求转发到与该请求对应的区域中的 DNS 服务器进行转换。DNS 具有可扩展性，这是因为主机名解析分散于多台服务器上完成。

不同的顶级域有不同的含义，分别代表着组织类型或起源国家/地区。常见域名后缀的含义如表 3.9 所示。

表 3.9　常见域名后缀的含义

序号	顶级域	含义
1	.net	.net 是国际最早被使用，也是最广泛流行的通用域名之一
2	.edu	edu 为 education 的简写。一般表示教育机构，比如大学等
3	.com	com 为 company 简称，表示公司企业。是目前国际最广泛流行的通用域名后缀之一
4	.org	适用于各类组织机构，包括非营利团体
5	.cn	.cn 是由我国管理的国际顶级域名，是中国自己的互联网标识。cn 代表中国，它体现了一种文化的认同、自身的价值和定位
6	.gov.cn	gov 是 government 的简写。.gov.cn 域名是中国政府机关等政府部门网站的重要标识，专门用于我国政府机关等部门

我们通常在配置网络设备时提供一个或者多个 DNS 服务器地址，DNS 客户端可以使用该地址进行域名解析。ISP 往往会为 DNS 服务器提供地址。当用户应用程序请求通过域名连入远程设备时，DNS 客户端将向某一域名服务器请求查询，获得域名解析后的数字地址。

用户还可以使用操作系统中名为 nslookup 的实用程序手动查询域名服务器，来解析给定的主机名。该实用程序也可以用于检修域名解析故障，以及验证域名服务器的当前状态，参考命令如下：

```
C:\Users\Administrator>nslookup
默认服务器: UnKnown
Address: fe80::1
> www.baidu.com
服务器: UnKnown
Address: fe80::1
非权威应答:
名称: www.a.shifen.com
Addresses: 183.232.231.174
          183.232.231.172
Aliases: www.baidu.com
>
```

DNS 查询的结果显示，域名 www.baidu.com 映射的 IP 地址为 183.232.231.174 和 183.232.231.172。

思考与训练

按如图3.48所示的项目拓扑图,在项目2思考与训练第7题的基础上继续完成以下需求。

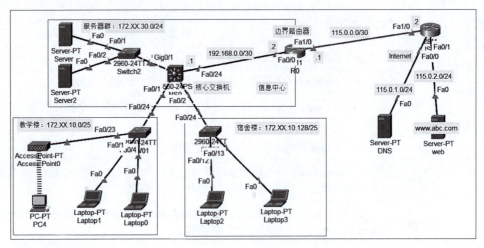

图3.48 项目拓扑图

(1)规划内网各部门的IP地址,适当配置,实现教学楼和宿舍楼的用户都能访问到学校的服务器。

(2)在校园内网配置静态路由,实现校园内全网互通(提示:所有用户应能访问到边界路由器的内网端口)。

(3)在内网的三层设备中配置默认路由,将发往Internet的数据包发到ISP。

(4)在边界路由器R0中应用PAT技术,实现内网所有用户能访问到Internet的服务器。

(5)在边界路由器R0中应用静态NAT技术,实现外部服务器能通过浏览器访问到内部两台服务器的Web界面。

(6)在Internet的DNS服务器添加www.abc.com的记录,实现内网用户可以通过浏览器,使用域名访问外网Web服务器。

注意:没有任何权限去配置Internet上的任何设备,包括ISP的路由器。

Project 3

Access the Internet

Users who access the network not only need to access local area network resources, but also have a greater need to access endless resources on the Internet. The goal of this project is to address the issue of user computers' accessing the Internet. This article mainly introduces the core technologies required for enterprise networks to access the Internet through ISP dedicated lines, such as static default routing, address translation, and domain name services.

Learning Objectives

(1) Understand how the internal network of the organization connects to the Internet and provides Internet access services for internal network users;
(2) Understand the role of static routing and cultivate the thinking of data routing;
(3) Understand the background of network address translation and its working mechanism;
(4) Understand the background and role of default routing;
(5) Understand the process of users' accessing web pages through domain names.

Skill Objectives

(1) Learn to apply static routing to achieve packet forwarding;
(2) Learn to apply PAT technology in boundary routers to enable internal network users to access the external network (Internet);
(3) Learn to apply static NAT technology in boundary routers to enable external users to access internal network resources;
(4) Master the method of summarizing routes and learn to flexibly apply default routes.

Relevant Knowledge

The majority of users accessing the Internet require access to the Internet, as the resources there are almost endless. So how can users access the Internet?

It is the operators (such as China Telecom, China Mobile, China Unicom, etc.) who provide us with accesses to the Internet. We need to send data to a server on the Internet to access its resources. First, we need to send this data to a certain operator, and then the operator sends the data to the destination server through its complex network (which may span across other operators).

1. Using Fiber Optics to Connect Enterprise Networks to Operator Networks

So the first thing we need to do is to choose a carrier you trust, and then connect the company's network to the carrier's network, see Figure 3.1.

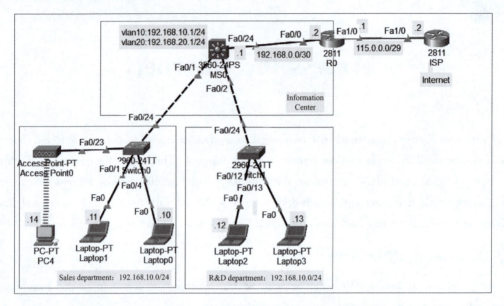

Figure 3.1　Enterprise Network Access to Operator Network

The company's internal network is connected to the operator's network through router R0. Because one side of this router is connected to the internal network and the other side is connected to the external network, this location happens to be the boundary between the internal and external networks, so it is commonly referred to as a boundary router.

Boundary routers are generally routers that connect to the operator's computer room through fiber optics. The boundary router is located in the company's data center and is usually very far from the operator's data center, so the fiber optic access is generally used. This requires both the border router and the operator's router to install optical modules. The installation of optical modules in the Cisco Packet Tracer simulator is shown in Figure 3.2.

Figure 3.2　Installation of optical modules in the router

The position of optical fibers in the media list is shown in Figure 3.3.

Figure 3.3　Media List in Cisco Packet Tracer

2. Configure a Direct Link Between the Border Router and the ISP Router

The planned network IP addresses between core switch MS0 and border router R0, and between R0 and the ISP's router, are shown in Table 3.1.

Table 3.1 IP Address Allocation Table

Equipment name	Port name	IP address	Subnet mask	Note
MS0	Fa0/24	192.168.0.1	255.255.255.252	extranet
R0	Fa0/0	192.168.0.2	255.255.255.252	
	Fa1/0	115.0.0.1	255.255.255.248	extranet
ISP	Fa1/0	115.0.0.2	255.255.255.248	

Configure the IP addresses for the corresponding devices according to the above planning with the following commands:

Configure the port (Fa0/24) of Layer 3 switch MS0 to connect to the border router.

```
MS0>en
MS0#conf t
MS0(config)#int fa 0/24
MS0(config-if)#no switchport
MS0(config-if)#ip add 192.168.0.1 255.255.255.252
MS0(config-if)#exit
MS0(config)#
```

Configure the hostname of border router R0 and its port.

```
Router>en
Router#conf t
Router(config)#hostname R0
R0(config)#int f 0/0
R0(config-if)#ip add 192.168.0.2 255.255.255.252
R0(config-if)#no sh
R0(config-if)#exit
R0(config)#int fa 1/0
R0(config-if)#ip add 115.0.0.1 255.255.255.248
R0(config-if)#no sh
R0(config-if)#exit
R0(config)#
```

Configure the carrier ISP router.

```
Router>en
Router#conf t
Router(config)#hostname ISP
ISP(config)#int fa 1/0
ISP(config-if)#no sh
ISP(config-if)#ip add 115.0.0.2 255.255.255.248
ISP(config-if)#
```

3. Testing of Connectivity Between Border Routers and Directly Connected Devices（MS0 and ISP routers） on the Internal and External Networks

Test the connectivity of the MS0→R0→ISP direct link by pinging the core switch MS0 and the ISP is router from R0 respectively, and the result should be through, as shown in Figure 3.4.

```
R0#ping 192.168.0.1
Type escape sequence to abort.
Sending 5, 100-byte ICMP Echos to 192.168.0.1, timeout is 2 seconds:
!!!!!
Success rate is 100 percent (5/5), round-trip min/avg/max = 0/0/0 ms
R0#ping 115.0.0.2
Type escape sequence to abort.
Sending 5, 100-byte ICMP Echos to 115.0.0.2, timeout is 2 seconds:
!!!!!
Success rate is 100 percent (5/5), round-trip min/avg/max = 0/2/8 ms
R0#
```

Figure 3.4　Test Results

Executing a ping command in a network device returns different information rather than executing a ping command on a PC. Routers use "!" (exclamation mark) to indicate normal communication, and "." (dots) for unexplained packet loss. The statistic "Success rate is 100 percent (5/5)" means that 5 packets have been tested and the success rate is 100 percent.

Our goal for this project is to enable users to access the Internet. That is, to allow intranet users to ping through the ISP router. The idea of implementation is: first of all, to let the intranet users send data to the border router, so that the border router helps forward to the ISP; ISP receives the data packet, and then gives the sender a confirmation of the answer packet to complete the whole communication process.

So the first problem that needs to be solved here is how users of the corporate intranet can send data to the border router. If the user can not send the data to the border router, then the border router can not receive the user's data, also can not help the user to forward the packet to the Internet. next we use static routing technology to solve this problem.

Task 1: Send Data to Border Routers—Static Routes

Task objective: to achieve user access to the border router

The example of Laptop3 sending data to a border router is analysed here. Laptop3 encapsulates an IP packet with the source and destination IP addresses shown in Table 3. 2.

Send Data to Border Routers— Static Routes

Table 3.2 Packet Address Information Sent by Laptop3 to the Border Router

Source IP address	Destination IP address	Network address of the destination network
192. 168. 20. 13/24	192. 168. 0. 2/30	192. 168. 0. 0/30

1. Analyse the Communication Process of IP Packets Sent by Intranet Users

According to the network topology of this project, Laptop3 (192. 168. 20. 13/24) and the border router (192. 168. 0. 2/30) are devices in different networks. According to the analysis in Item 2, the devices in different networks must go through the gateway to help forward the data, so Laptop3 needs to send the data to its own gateway (192. 168. 20. 1/24) first, and then the Layer 3 switch MS0 will query the routing table, and make the forwarding decision based on the route entries in the routing table: To discard or to forward? From which port to forward out?

We can already determine that Laptop3 can ping the gateway in Project 2, so we will not repeat the test here. After the gateway gets the data that Laptop3 sends to the border router R0, in order to determine how the data is forwarded, MS0 needs to look up the routing table based on the destination address of the packet to be forwarded. The routing table of MS0 is shown in Figure 3. 5.

```
MS0#show ip route
Codes: C - connected, S - static, I - IGRP, R - RIP, M - mobile, B - BGP
       D - EIGRP, EX - EIGRP external, O - OSPF, IA - OSPF inter area
       N1 - OSPF NSSA external type 1, N2 - OSPF NSSA external type 2
       E1 - OSPF external type 1, E2 - OSPF external type 2, E - EGP
       i - IS-IS, L1 - IS-IS level-1, L2 - IS-IS level-2, ia - IS-IS inter
area
       * - candidate default, U - per-user static route, o - ODR
       P - periodic downloaded static route

Gateway of last resort is not set

     192.168.0.0/30 is subnetted, 1 subnets
C       192.168.0.0 is directly connected, FastEthernet0/24
C    192.168.10.0/24 is directly connected, Vlan10
C    192.168.20.0/24 is directly connected, Vlan20
```

Figure 3.5 Routing Table for Core Router MS0

In the routing table, MS0 finds a routing entry for the destination network (192. 168. 0. 0). This routing entry tells the router that network 192. 168. 0. 0 is then directly connected (directly connected) to port FastEthernet0/24. If there is data to be sent to network 192. 168. 0. 0, the data is forwarded out of port FastEthernet0/24.

According to the topology shown in Figure 3. 1, R0 can receive the data after MS0 forwards it out of port FastEthernet0/24. The whole process is smooth and does not require any additional configuration.

2. Analyse the Process by Which Border Router R0 Sends An Answer Packet to An Intranet User

We define a smooth information transfer process as-data has a way to go and a way to come back. At the moment it is only certain that the data sent by Laptop3 can reach the border router R0 and that the border router can receive the packet. But when R0 wants to send an answer packet back to Laptop3, the border router R0 encapsulates an IP packet with the source and destination addresses shown in Table 3.3.

Table 3.3 Packet Address Information Sent by the Border Router to Laptop3

Source IP address	Destination IP address	Network address of the destination network
192.168.0.2/30	192.168.20.13/24	192.168.20.0/24

R0 also needs to look up its routing table based on the destination address of the packet (192.168.20.13/24) in order to decide what to do with the packet: should it be discarded or forwarded? From which port will the answer packet go out?

The routing table for border router R0 is shown in Figure 3.6.

```
R0#sh ip rou
Codes: C - connected, S - static, I - IGRP, R - RIP, M - mobile, B - BGP
       D - EIGRP, EX - EIGRP external, O - OSPF, IA - OSPF inter area
       N1 - OSPF NSSA external type 1, N2 - OSPF NSSA external type 2
       E1 - OSPF external type 1, E2 - OSPF external type 2, E - EGP
       i - IS-IS, L1 - IS-IS level-1, L2 - IS-IS level-2, ia - IS-IS inter area
       * - candidate default, U - per-user static route, o - ODR
       P - periodic downloaded static route

Gateway of last resort is not set

     115.0.0.0/29 is subnetted, 1 subnets
C       115.0.0.0 is directly connected, FastEthernet1/0
     192.168.0.0/30 is subnetted, 1 subnets
C       192.168.0.0 is directly connected, FastEthernet0/0
R0#
```

Figure 3.6 Routing Table for R0

R0 cannot find the network address (192.168.20.0/24) of Laptop3 in the routing table, and according to the rules, it can only discard the answer packet, which leads to the interruption of communication.

One simple solution to this is to manually add a record for the 192.168.20.0 network to the routing table so that the router can make forwarding decisions by querying the routing table.

3. Adding Static Routes to the Routing Table

Manually adding a piece of routing information to the routing table, this routing information is called a static route. Nert we will add a piece of information to R0's routing table that tells the router to forward data to MS0 (192.168.0.1) if there is data sent to the 192.168.20.0/24 network. The command is shown in Figure 3.7.

```
R0#conf t
R0(config)#ip route 192.168.20.0 255.255.255.0 192.168.0.1
```

Figure 3.7 Static Route Commands

Static routing commands explained:

①The "IP route" command adds a static route;

②"192.168.20.0 255.255.255.0" indicates the network number and subnet mask of the final destination network;

③"192.168.0.1" indicates the next hop address.

The whole command means that if there is data to be sent to the network 192.168.20.0/24, the data is forwarded to 192.168.0.1, which is responsible for the next forwarding.

Examining the routing table with respect to the routing entry for the 192.168.20.0 network resulted in the results shown in Figure 3.8.

```
R0#sh ip rou
Codes: C - connected, S - static, I - IGRP, R - RIP, M - mobile, B - BGP
       D - EIGRP, EX - EIGRP external, O - OSPF, IA - OSPF inter area
       N1 - OSPF NSSA external type 1, N2 - OSPF NSSA external type 2
       E1 - OSPF external type 1, E2 - OSPF external type 2, E - EGP
       i - IS-IS, L1 - IS-IS level-1, L2 - IS-IS level-2, ia - IS-IS inter area
       * - candidate default, U - per-user static route, o - ODR
       P - periodic downloaded static route

Gateway of last resort is not set

     115.0.0.0/29 is subnetted, 1 subnets
C       115.0.0.0 is directly connected, FastEthernet1/0
     192.168.0.0/30 is subnetted, 1 subnets
C       192.168.0.0 is directly connected, FastEthernet0/0
S    192.168.20.0/24 [1/0] via 192.168.0.1
```

Figure 3.8 Information about Newly Added Route Entries in the R0 Routing Table

According to the output routing table, there is an additional route 192.168.20.0 labelled "S" in the table. According to the explanation above, "S" stands for "static". This means that the route to the 192.168.20.0/24 network has been successfully added, and if you check the routing table after R0, you will know that if you want to send data to the 192.168.20.0/24 network, you will send the data to 192.168.0.1.

In fact, R0 does not know who 192.168.0.1 is at this time, from which port to forward out. So R0 will check the routing table again (recursive query), and find the record shown in Figure 3.9 in order to be sure: the data to 192.168.0.1 is sent out from port FastEthernet0/0.

```
R0#sh ip rou
Codes: C - connected, S - static, I - IGRP, R - RIP, M - mobile, B - BGP
       D - EIGRP, EX - EIGRP external, O - OSPF, IA - OSPF inter area
       N1 - OSPF NSSA external type 1, N2 - OSPF NSSA external type 2
       E1 - OSPF external type 1, E2 - OSPF external type 2, E - EGP
       i - IS-IS, L1 - IS-IS level-1, L2 - IS-IS level-2, ia - IS-IS inter area
       * - candidate default, U - per-user static route, o - ODR
       P - periodic downloaded static route

Gateway of last resort is not set

     115.0.0.0/29 is subnetted, 1 subnets
C       115.0.0.0 is directly connected, FastEthernet1/0
     192.168.0.0/30 is subnetted, 1 subnets
C       192.168.0.0 is directly connected, FastEthernet0/0
S    192.168.20.0/24 [1/0] via 192.168.0.1
```

Figure 3.9 Route Entry Information in the R0 Routing Table

4. Continue to Analyse the Communication Process for Data Return

After the data arrives at 192.168.0.1 (MS0), MS0 also goes to its own routing table, and according to the routing table, as shown in Figure 3.10, it finds out information about the 192.168.20.0/24 network, and finally forwards the data out of port VLAN 20.

```
MS0#show ip route
Codes: C - connected, S - static, I - IGRP, R - RIP, M - mobile, B - BGP
       D - EIGRP, EX - EIGRP external, O - OSPF, IA - OSPF inter area
       N1 - OSPF NSSA external type 1, N2 - OSPF NSSA external type 2
       E1 - OSPF external type 1, E2 - OSPF external type 2, E - EGP
       i - IS-IS, L1 - IS-IS level-1, L2 - IS-IS level-2, ia - IS-IS inter
area
       * - candidate default, U - per-user static route, o - ODR
       P - periodic downloaded static route

Gateway of last resort is not set

     192.168.0.0/30 is subnetted, 1 subnets
C       192.168.0.0 is directly connected, FastEthernet0/24
C    192.168.10.0/24 is directly connected, Vlan10
C    192.168.20.0/24 is directly connected, Vlan20
```

Figure 3.10 Routing Information in the MS0 Routing Table

After Laptop3 gets the answer packet, the communication is complete. The result of Laptop3 host pinging border router R0 is shown in Figure 3.11.

```
C:\>ping 192.168.0.2

Pinging 192.168.0.2 with 32 bytes of data:

Reply from 192.168.0.2: bytes=32 time<1ms TTL=254
Reply from 192.168.0.2: bytes=32 time=3ms TTL=254
Reply from 192.168.0.2: bytes=32 time=7ms TTL=254
Reply from 192.168.0.2: bytes=32 time=1ms TTL=254

Ping statistics for 192.168.0.2:
    Packets: Sent = 4, Received = 4, Lost = 0 (0% loss),
Approximate round trip times in milli-seconds:
    Minimum = 0ms, Maximum = 7ms, Average = 2ms
```

Figure 3.11 Test Results

5. Add a Static Route to R0 to Enable VLAN 10 Users to Access the Border Router

How to enable the hosts in the sales department network to ping the border router? Analysing it by the same way of thinking, you will find that the data is able to reach the border router R0, it is just that R0 does not know from which port the answer packet should be sent back.

You can add a static route to R0 that tells R0: If there is data to be sent to network 192.168.10.0/24, please forward it to 192.168.0.1. The command is as follows:

R0>en

R0#conf t

R0(config)#ip route 192.168.10.0 255.255.255.0 192.168.0.1 //
Configure a static route that sends data to the 192.168.10.0/24 network to node 192.168.0.1, which takes care of the next forwarding.

Check the routing table to make sure it is added successfully, as shown in Figure 3.12.

```
R0#sh ip rou
Codes: C - connected, S - static, I - IGRP, R - RIP, M - mobile, B - BGP
       D - EIGRP, EX - EIGRP external, O - OSPF, IA - OSPF inter area
       N1 - OSPF NSSA external type 1, N2 - OSPF NSSA external type 2
       E1 - OSPF external type 1, E2 - OSPF external type 2, E - EGP
       i - IS-IS, L1 - IS-IS level-1, L2 - IS-IS level-2, ia - IS-IS inter area
       * - candidate default, U - per-user static route, o - ODR
       P - periodic downloaded static route

Gateway of last resort is not set

     115.0.0.0/29 is subnetted, 1 subnets
C       115.0.0.0 is directly connected, FastEthernet1/0
     192.168.0.0/30 is subnetted, 1 subnets
C       192.168.0.0 is directly connected, FastEthernet0/0
S    192.168.10.0/24 [1/0] via 192.168.0.1
S    192.168.20.0/24 [1/0] via 192.168.0.1
```

Figure 3.12   Routing Table for R0

After confirming the successful addition of the routing table, test pinging the border router (192.168.0.2) from Laptop0 with the results shown in Figure 3.13.

```
C:\>ping 192.168.0.2

Pinging 192.168.0.2 with 32 bytes of data:

Reply from 192.168.0.2: bytes=32 time=1ms TTL=254
Reply from 192.168.0.2: bytes=32 time<1ms TTL=254
Reply from 192.168.0.2: bytes=32 time=1ms TTL=254
Reply from 192.168.0.2: bytes=32 time=1ms TTL=254
```

Figure 3.13   Test Results

At this point, the intranet users can access the border router, the enterprise network can be all internally inter-connected.

Mastering static routing techniques allows you to control the flow of data at will. Whoever you want to send the data to, point the next hop of the static route. Next, if there is data that needs to be sent to the Internet, the internal hosts will first send the data to the border router, and then the border router R0 will forward the data to the ISP's network, and the same can be done with static routing to send the intranet data towards the Internet—the next task will analyse and achieve this goal.

Exercise: Referring to this task, configure a static route so that all intranet computers can access the border router R0 based on the enterprise network model.

Competence Expansion: Routers decide how to forward packets based on the routing table, so the routing entries in the routing table are particularly important. Once the routing table is matched successfully, the router strictly forwards the data to the next hop node recorded in the routing entry that has been matched. In a way, the routing table represents a flag that determines the direction of data flow, so it is especially important to keep the routing entries in the routing table correct. College students should also have a flag inside of their hearts, a flag of patriotism, to unite their love for the country, the will to strengthen the country, and the act of serving the country together, so as to make due contributions to the country and the nation.

Task 2: Enabling Intranet Users to Access Internet Servers—Port Address Translation (PAT)

Enabling Intranet Users to Access Internet Servers— Port Address Translation(PAT)

Mission Objective: Translate the intranet address at the border router to enable intranet users to access the extranet (ping through the ISP router address: 115.0.0.2/29).

After the configuration of the previous tasks, will the computers on the corporate intranet be able to access the extranet without any problem? The test has found that the intranet users are still unable to access the extranet host. In order to find out the cause of the problem, we will analyse the whole process of data communication from the sending and the returning of data.

Continuing with the analysis of the network topology of this project, the Laptop3 has to ping the address of the operator 115.0.0.2/29. The Laptop3 encapsulates an IP packet with the source and destination addresses as shown in Table 3.4.

Table 3.4 Packet Address Information Sent by Laptop3 to ISP Router

Source IP address	Destination IP address	Network address of the destination network
192.168.20.13/24	115.0.0.2/29	115.0.0.0/29

Will the packets sent by Laptop3 reach the ISP router without any problem? We find that as long as the Layer 3 device ensures that there is a routing entry for the destination network, there should be no problem with data forwarding.

1. Analysis of IP Packets Sent From Laptop3

Laptop3 discovers that the destination address 115.0.0.2/29 is on a different network, so it sends the IP packet to its gateway (MS0: 192.168.20.1/24) to ask for help in forwarding it. When MS0 receives the IP packet, it queries the routing table based on the destination IP address on the IP packet, as shown in Figure 3.14.

```
MS0#sh ip route
Codes: C - connected, S - static, I - IGRP, R - RIP, M - mobile, B - BGP
       D - EIGRP, EX - EIGRP external, O - OSPF, IA - OSPF inter area
       N1 - OSPF NSSA external type 1, N2 - OSPF NSSA external type 2
       E1 - OSPF external type 1, E2 - OSPF external type 2, E - EGP
       i - IS-IS, L1 - IS-IS level-1, L2 - IS-IS level-2, ia - IS-IS inter area
       * - candidate default, U - per-user static route, o - ODR
       P - periodic downloaded static route

Gateway of last resort is not set

     192.168.0.0/30 is subnetted, 1 subnets
C       192.168.0.0 is directly connected, FastEthernet0/24
C    192.168.10.0/24 is directly connected, Vlan10
C    192.168.20.0/24 is directly connected, Vlan20
```

Figure 3.14 MS0 Routing Table

MS0 discovers that there is no route entry in the routing table for the destination network (115.0.0.0/29), which by rule can only be discarded. One of the solutions is to add static routes to the routing table along the lines of Task 1 of this project.

2. Add Static Route to MS0 to Forward Data Destined for ISP Router to Border Router R0

Add a routing entry for 115. 0. 0. 0/29 to MS0 with the following command:

```
MS0(config)#ip route 115.0.0.0 255.255.255.248 192.168.0.2
```

Check the routing table to confirm that it has been added successfully, as shown in Figure 3. 15.

```
MS0#sh ip rou
Codes: C - connected, S - static, I - IGRP, R - RIP, M - mobile, B - BGP
       D - EIGRP, EX - EIGRP external, O - OSPF, IA - OSPF inter area
       N1 - OSPF NSSA external type 1, N2 - OSPF NSSA external type 2
       E1 - OSPF external type 1, E2 - OSPF external type 2, E - EGP
       i - IS-IS, L1 - IS-IS level-1, L2 - IS-IS level-2, ia - IS-IS inter area
       * - candidate default, U - per-user static route, o - ODR
       P - periodic downloaded static route

Gateway of last resort is not set

     115.0.0.0/29 is subnetted, 1 subnets
S       115.0.0.0 [1/0] via 192.168.0.2
     192.168.0.0/30 is subnetted, 1 subnets
C       192.168.0.0 is directly connected, FastEthernet0/24
C    192.168.10.0/24 is directly connected, Vlan10
C    192.168.20.0/24 is directly connected, Vlan20
```

Figure 3. 15 Routing Table for Core Switch MS0

MS0 forwards the data to 192. 168. 0. 2 (border router R0) based on the newly-added route entry. R0 likewise goes to the routing table, as shown in Figure 3. 16.

```
R0#sh ip rou
Codes: C - connected, S - static, I - IGRP, R - RIP, M - mobile, B - BGP
       D - EIGRP, EX - EIGRP external, O - OSPF, IA - OSPF inter area
       N1 - OSPF NSSA external type 1, N2 - OSPF NSSA external type 2
       E1 - OSPF external type 1, E2 - OSPF external type 2, E - EGP
       i - IS-IS, L1 - IS-IS level-1, L2 - IS-IS level-2, ia - IS-IS inter area
       * - candidate default, U - per-user static route, o - ODR
       P - periodic downloaded static route

Gateway of last resort is not set

     115.0.0.0/29 is subnetted, 1 subnets
C       115.0.0.0 is directly connected, FastEthernet1/0
     192.168.0.0/30 is subnetted, 1 subnets
C       192.168.0.0 is directly connected, FastEthernet0/0
S    192.168.10.0/24 [1/0] via 192.168.0.1
S    192.168.20.0/24 [1/0] via 192.168.0.1
```

Figure 3. 16 Routing Table for Border Router R0

The routing table tells R0 to send to the network at 115. 0. 0. 0/29 out of the FastEthernet1/0 port. Eventually the data reaches the opposite end-the ISP's router.

Obviously, by adding a route entry to the intranet, the data can be sent to the destination network on the extranet without any problem. But can the destination network host (ISP router) on the external network return the answer packet to the sender Laptop3 without any problem? Let's continue to analyse the returning of the answer packet.

3. Analyse the Process of Returning An Answer Packet from the ISP Router

After the ISP's router receives the packet, it generates an answer packet to send back to Laptop3 with the address information shown in Table 3.5.

Table 3.5 Packet Address Information Sent by the ISP Router to Laptop3

Source IP address	Destination IP address	Network address of the destination network
115.0.0.2/29	192.168.20.13/24	192.168.20.0/24

In order to send this packet back to Laptop3, the ISP router goes to its own routing table in order to determine which port to send the packet from. The ISP's routing table is shown in Figure 3.17.

```
ISP#sh ip rou
Codes: C - connected, S - static, I - IGRP, R - RIP, M - mobile, B - BGP
       D - EIGRP, EX - EIGRP external, O - OSPF, IA - OSPF inter area
       N1 - OSPF NSSA external type 1, N2 - OSPF NSSA external type 2
       E1 - OSPF external type 1, E2 - OSPF external type 2, E - EGP
       i - IS-IS, L1 - IS-IS level-1, L2 - IS-IS level-2, ia - IS-IS inter area
       * - candidate default, U - per-user static route, o - ODR
       P - periodic downloaded static route

Gateway of last resort is not set

     115.0.0.0/29 is subnetted, 1 subnets
C       115.0.0.0 is directly connected, FastEthernet1/0
```

Figure 3.17 Routing Table of an ISP Router

However, from the output routing information, the ISP router's routing table does not have a routing entry for the 192.168.20.0/24 network. As a rule, the ISP will drop the answer packet, resulting in discontinuity of data communication.

Is it possible to add a static route to the ISP's router and have the router send the packets back to Laptop3? Many people, after mastering the static routing technique, will easily think of following the idea of Task 1 of this project, adding a static route to the ISP router and letting the ISP just send the data back to R0.

But this creates two problems:

①Private IP addresses cannot be routed on the Internet because the routing tables of public network devices do not allow records of private IP addresses to exist;

②Private network addresses are planned within the organisation itself and can be used at will. Multiple customers of an operator who use the same private network address, if the private network address is allowed to appear on the public network's routers, the operator must configure the routing entries to reach the subscriber's network, and for a given network that is being reused, the next-hop address is not unique, which ultimately leads to data loss. For example, if Company A and Company B are both connected to the same carrier, and both companies use the same private IP address segment (192.168.0.0/24), should the carrier point to Company A or Company B for the next-hop node that reaches the 192.168.0.0/24 network when it writes the routing entry? If pointing to company A, then company B will not receive the return traffic; if pointing to company B, company A will not receive the return traffic either; if you write two, one pointing to company A, the other pointing to company B, then the two companies A and B will only receive 50% of the return traffic. Obviously neither will work.

The solution to the problem is the application of Network Address Translation (NAT) technology. The border router is responsible for converting the source IP address (private address) of the packets sent by Laptop3 into a public IP address. This is because when the ISP's router generates an answer packet, the destination address is determined based on the source address of the packet it receives (similar to when we receive a letter, and when we reply to the letter, the destination address is written as the source address of the received letter). In this way, when the ISP's router returns the answer packet, the destination address is also the public address, avoiding the illegal existence of private addresses on the public network.

PAT (Port Address Translation) is one of the most commonly-used methods of NAT (Network Address Translation) technology. In the following part, the PAT technique is applied to the border router to transform the source address of the packet (private IP address to public IP address).

Step 1: Define NAT Intranet Ports and Extranet Ports.

Against the project topology, port Fa0/0 of border router R0 is connected to the corporate intranet, as shown in Figure 3.18.

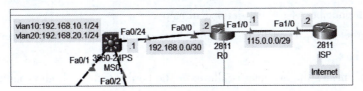

Figure 3.18 Project Topology

So here you need to define this port as an intranet port for NAT. The command is as follows:

```
R0>
R0>en
R0#conf t
R0(config)#int fa 0/0
R0(config-if)#ip nat inside   //define port Fa0/0 as a NAT intranet port
R0(config-if)#exit
R0(config)#
```

The Fa1/0 port of router R0 is connected to the corporate intranet, so here you need to define this port as an external port for NAT. The command is as follows:

```
R0(config)#int fa 1/0
R0(config-if)#ip nat outside   //Define port Fa1/0 as a NAT external port
R0(config-if)#exit
```

Step 2: Capture the Traffic to be Converted with Access Control Lists (ACLs). To Capture the Packets Sent from the Network Where Laptop3 is Located, the Command is as follows:

```
R0(config)#access-list 1 permit 192.168.20.0 0.0.0.255
```

Here a list is created, the number is "1", this list can capture packets sent from network 192.168.20.0/24. (The number can be defined according to the situation, but to comply with the rules)

Note: Wildcard bits are used here instead of the previously used mask format. One way to calculate a wildcard mask is to simply subtract the subnet mask from 255.255.255.255. In this example, 255.255.255.255 minus 255.255.255.0 gives 0.0.0.255.

Step 3: Map the Captured Traffic to An External Port. The Command is as Follows.

```
R0(config)#ip nat inside source list 1 interface fastEthernet 1/0 overload
```

Parameter Interpretation:

①The "ip nat" command asks the router to perform address translation.

②The "inside source" parameter indicates that the source address of the packet needs to be converted when the packet is sent from the internal network to the external network.

③"list 1" means the access control list numbered "1". Note that the number of the list should be the same as the number of the list defined in step 2.

④"fastEthernet 1/0" refers to the external port of border router R0, which connects to the carrier's network port.

⑤"overload" means the overload is allowed. The border router will allow multiple users in the intranet to implement address translation at the same time by means of port mapping. The theoretical number of address translations can be up to 64,511 (65535-1024).

Step 4: Test Access to the Extranet from An Intranet Host.

The result of pinging the external ISP router (115.0.0.2/29) from Laptop3 is shown in Figure 3.19.

```
C:\>ping 115.0.0.2

Pinging 115.0.0.2 with 32 bytes of data:

Reply from 115.0.0.2: bytes=32 time<1ms TTL=253
Reply from 115.0.0.2: bytes=32 time=3ms TTL=253
Reply from 115.0.0.2: bytes=32 time=11ms TTL=253
Reply from 115.0.0.2: bytes=32 time=13ms TTL=253
```

Figure 3.19　Laptop3 Ping External ISP Router (115.0.0.2/29) Result

The result shows that the communication is normal and we have successfully achieved the access of the intranet users to the extranet by applying the PAT technique. At border router R0 we can view the details of address translation as shown in Figure 3.20.

```
R0#sh ip nat translations
Pro   Inside global      Inside local       Outside local      Outside global
icmp  115.0.0.1:17       192.168.20.13:17   115.0.0.2:17       115.0.0.2:17
icmp  115.0.0.1:18       192.168.20.13:18   115.0.0.2:18       115.0.0.2:18
icmp  115.0.0.1:19       192.168.20.13:19   115.0.0.2:19       115.0.0.2:19
icmp  115.0.0.1:20       192.168.20.13:20   115.0.0.2:20       115.0.0.2:20
```

Figure 3.20　Address Translation Information for a Border Router

The show ip nat translations command displays information about the translation, each packet translating the addresses of four icmp packets. These include the source address (192.168.20.13) to the internal global address (Inside global: 115.0.0.1) when the packet goes out of the intranet; and the destination address (115.0.0.1) to the internal local address (192.168.20.13) when the packet enters the intranet from the outside network. Round-trip packets undergo the opposite process of address translation.

The captured packet statistics can also be viewed through the ACL, and a total of eight packets are captured, as shown in Figure 3.21.

```
R0#
R0#sh ip access-lists 1
Standard IP access list 1
    permit 192.168.20.0 0.0.0.255 (8 match(es))
R0#
```

Figure 3.21 Examining Captured Packets on Border Router R0

The datagram statistics captured by the border router access control list are "8 match(es)": 8 packets are matched. The ping command only sends out 4 packets, and the show ip translations command only shows 4 records. However, the captured statistics include the 4 packets that go out and the 4 packets that come back. (Note: Both outgoing and returning packets need to be translated, except that when the packets go out, the source address is private and needs to be converted to public; when the packets come back, the destination address is converted: the destination address is public and needs to be converted to the original private address.)

At this time, the users on VLAN 10 in the sales department still can't access the outside network. Because the border router R0 is not capturing the traffic from their network, it is not translating the addresses of the packets. As analysed at the beginning of this task, this results in the packets being dropped when they reach their destination because the destination address of the answer packet is a private IP address and could not be routed back.

The solution is simple adding a rule to the access control list of the border router to allow the router to capture the data sent from VLAN 10.

Step 5: Add a Rule to the Access Control List of Border Router R0 to Allow the Router to Capture the Data Sent From the VLAN 10 Network. The Command is As Follows:

```
R0(config)#access-list 1 permit 192.168.10.0 0.0.0.255
```

Test the Sales Department user Laptop0 to access the extranet, and the result is shown in Figure 3.22.

```
C:\>ping 115.0.0.2

Pinging 115.0.0.2 with 32 bytes of data:

Reply from 115.0.0.2: bytes=32 time=1ms TTL=253
Reply from 115.0.0.2: bytes=32 time=14ms TTL=253
Reply from 115.0.0.2: bytes=32 time=15ms TTL=253
Reply from 115.0.0.2: bytes=32 time=12ms TTL=253
```

Figure 3.22 Laptop0 Access to ISP Router Answer Message

The results show that the computers of the users in the sales department can also access the extranet normally.

The results show that the computers of the users in the sales department also have normal access to the extranet.

Exercise: Referring to this task, configure PAT on the border road router so that all users on the corporate intranet can access the ISP's router. Note: The ISP router cannot have any route from the intranet on it.

Competence Expansion: Operators provide users with access to the Internet. People can obtain information through the Internet in a more convenient and diversified way, and most people, especially young people, rely more and more mainly on the Internet to obtain information. At the same time, the Internet is also filled with more and more negative information content, which affects the order of online life. College students should use the Internet correctly, improve the ability to obtain information, strengthen the ability to identify information, and enhance the ability to apply information, so that the Internet can become an important tool to broaden their horizons and improve their abilities.

Task 3: Enabling Internet Users to Access Intranet Hosts—Network Address Translation (Static NAT)

Enabling Internet Users to Access Intranet Hosts— Network Address Translation (Static NAT)

Task Objective: Apply static NAT technology to enable the external network users to access the internal network host.

The last task we use PAT technology to enable the internal users to access the external host, if the external users want to access the internal host, how to achieve it? —There is a technology to achieve this demand, that is, static NAT technology: the one-to-one mapping of the private IP address and the public IP address one-to-one mapping.

Next, we configure static NAT technology to enable the ISP's router to ping Laptop3.

Step 1: Define NAT Intranet Ports and Extranet Ports.

According to the project topology, port Fa0/0 of border router R0 is connected to the corporate intranet, as shown in Figure 3.23.

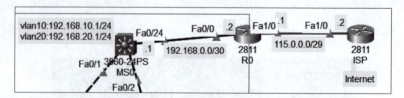

Figure 3.23　Project Topology

So here you need to define the Fa0/0 port of R0 as an intranet port for NAT. The command is as follows:

```
R0>
R0>enable
R0#conf t
R0(config)#int fa0/0
R0(config-if)#ip nat inside   //Define port Fa0/0 as a NAT internal port
R0(config-if)#exit
R0(config)#
```

The Fa1/0 port of Router R0 is connected to the corporate intranet, so here you need to define the port as an external port for NAT. The command is as follows:

```
R0(config)#int fa1/0
R0(config-if)#ip nat outside    //Define port Fa1/0 as a NAT external port
R0(config-if)#exit
```

Note: If you have already defined the NAT intranet port and extranet port in Task 2 of this project, you do not need to repeat the configuration and can skip this step.

Step 2: Mapping Private IP Addresses to Public IP Addresses.

In the global mode, the IP address of Laptop3 on the internal network (192.168.20.13) is mapped to an address on the network (115.0.0.0/29) where the external port is located.

After calculation, the range of host addresses for network 115.0.0.0/29 is 115.0.0.1/29 to 115.0.0.6/29. Since 115.0.0.1 and 115.0.0.2 are already occupied by the two devices, R0 and ISP, here we use the host address that is currently unused on that network: 115.0.0.3.

The commands to configure static NAT mapping are as follows:

```
R0(config)#ip nat inside source static 192.168.20.13 115.0.0.3
```

Order Interpretation:

①The "static" parameter means static.

②"192.168.20.13" is the IP address of the intranet host Laptop3.

③"115.0.0.3" is the legal public address on the same network segment as the external port of the border router. (The IP address of the external port can be used.)

In this way, users from the external network can access the public address 115.0.0.3, and then the border router forwards the data from the external network to the 192.168.20.13 host on the internal network. Thus, users from the external network can access the resources of the internal host.

Step 3: Test ISP Router Access to Laptop3.

Note that a direct ping of the Laptop3's IP address from the ISP router is unsuccessful, because there must be no routing entry for the private address on the external network. And at this point, the address of Laptop3 on the internal network (192.168.20.13) has been mapped to an address that can be recognised by the external network (115.0.0.3), so for the test, you should ping the mapped external address (115.0.0.3) from the external network, as shown in Figure 3.24.

```
ISP#ping 115.0.0.3

Type escape sequence to abort.
Sending 5, 100-byte ICMP Echos to 115.0.0.3, timeout is 2 seconds:
!!!!!
Success rate is 100 percent (5/5), round-trip min/avg/max = 0/13/28 ms
```

Figure 3.24 ISP Router Ping Laptop3 Host Mapped Address Results

The results show that communication is normal. Viewing the conversion logs at the border router, the results are shown in Figure 3.25.

```
R0#show ip nat translations
Pro   Inside global      Inside local       Outside local      Outside global
---   115.0.0.3          192.168.20.13      ---                ---

R0#
```

Figure 3.25 NAT Mapping Table for Border Routers

We turn on debug at the border router R0, as shown in Figure 3.26.

```
R0#debug ip nat
IP NAT debugging is on
R0#
```

Figure 3.26 Enabling Debug in R0

At the ISP router, ping the address mapped by host Laptop3 again, and then view the debug message at router R0 as shown in Figure 3.27.

```
R0#debug ip nat
IP NAT debugging is on
R0#
NAT: s=115.0.0.2, d=115.0.0.3->192.168.20.13 [47]

NAT*: s=192.168.20.13->115.0.0.3, d=115.0.0.2 [40]

NAT: s=115.0.0.2, d=115.0.0.3->192.168.20.13 [48]

NAT*: s=192.168.20.13->115.0.0.3, d=115.0.0.2 [41]

NAT: s=115.0.0.2, d=115.0.0.3->192.168.20.13 [49]

NAT*: s=192.168.20.13->115.0.0.3, d=115.0.0.2 [42]

NAT: s=115.0.0.2, d=115.0.0.3->192.168.20.13 [50]

NAT*: s=192.168.20.13->115.0.0.3, d=115.0.0.2 [43]

NAT: s=115.0.0.2, d=115.0.0.3->192.168.20.13 [51]

NAT*: s=192.168.20.13->115.0.0.3, d=115.0.0.2 [44]
```

Figure 3.27 R0 Router Debug Infographic

Information Interpretation:

①"NAT" indicates the conversion information of the packet when it enters the internal network from the external network.

②"NAT *" indicates the conversion information when the packet is sent out from the internal network to the external network.

③The d in "d=115.0.0.3->192.168.20.13" denotes the destination address, which means that when a packet enters the intranet from the extranet, the destination address of the packet, 115.0.0.3, is converted into the address 192.168.20.13. 192.168.20.13.

④The s in "s=192.168.20.13->115.0.0.3" denotes the source address, which means that when an answer packet is returned from the intranet to the extranet, it translates the source address 192.168.20.13 into 115.0.0.3.

⑤The final middle bracket [47] indicates that the port number used for NAT conversion is 47.

Static address mapping can be accurate interms of the port number. In this case, we are mapping two IP addresses to each other, and all port numbers for these two IP addresses are also one-to-one. Of course, if you just want to map a specific port number, that is also possible. The following command maps port 8080 on the intranet address (192.168.20.13) to port 80 on the extranet address (115.0.0.3) as follows:

```
R0(config)#ip nat inside source static tcp 192.168.20.13 8080 115.0.0.3 80
```

Exercise: Referring to this task, add a server to the corporate intranet, access the core switch, and configure it appropriately so that the server can access the border router. Then configure PAT on the border router R0 so that Internet users can access the server on the corporate intranet.

Competence Expansion: Static NAT mapping technology is usually applied to achieve the scenario in which servers deployed in the internal network can provide services to external users. However, this technique exposes internal servers to the Internet and great external threats. Engineers need to strictly follow the GB/T 22239—2019 "Information Security Technology-Network Security Level Protection Requirements" standard to build a strong fortress for information systems to prevent information leakage. College students need to build a strong fortress of their own thoughts, enhance their security awareness, maintain a high degree of vigilance against the infiltration, subversion, and destructive activities of hostile forces inside and outside the country, and effectively fulfil their obligations to safeguard national security.

Task 4: Reducing Route Entries in the Routing Table—Route Summarisation (Default Route)

Task Objective: Send all data from the intranet out to the extranet by using one default route.

In Task 1 of this project, in order to enable hosts on the intranet to access hosts on a network on the extranet, it is necessary to add routing entries to all Layer 3 devices on the intranet (core

Reducing Route Entries in the Routing Table— Route Summarisation (Default Route)

switches and border routers), or else the Layer 3 devices will drop the data if they don't find the relevant routing information for the destination network in their own routing tables.

In order to be able to better understand this knowledge point, let's fill in the ISP part of the project topology: add two servers from a carrier's Internet Data Center (IDC). This is shown in Figure 3.28.

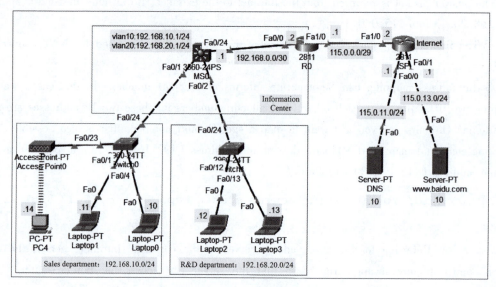

Figure 3.28 Task Topology

1. Configure the IP Address of the Server As Well As the Gateway so that Devices on the External Network Can Communicate With Each Other

The IP address plan for each device is shown in Table 3.6.

Table 3.6 Internet Data Centre IP Address Information Table

Equipment name	Ports	IP address	Note
DNS server		115.0.11.10/24	
web server		115.0.13.10/24	
ISP router	Fa0/0	115.0.11.1/24	DNS server gateway
	Fa0/1	115.0.13.1/24	Web server gateway

2. Add Routes to Layer 3 Devices on the Intranet to Send Data From the Intranet to ISP Data Centre Servers

In order to enable users on the intranet to access the two networks on the extranet: 115.0.11.0/24 and 115.0.13.0/24, we must add two routes to the core switch on the intranet so that the core switch can forward data destined for these two networks to R0, as shown in Figure 3.29.

```
MS0#sh ip rou
Codes: C - connected, S - static, I - IGRP, R - RIP, M - mobile, B - BGP
       D - EIGRP, EX - EIGRP external, O - OSPF, IA - OSPF inter area
       N1 - OSPF NSSA external type 1, N2 - OSPF NSSA external type 2
       E1 - OSPF external type 1, E2 - OSPF external type 2, E - EGP
       i - IS-IS, L1 - IS-IS level-1, L2 - IS-IS level-2, ia - IS-IS inter area
       * - candidate default, U - per-user static route, o - ODR
       P - periodic downloaded static route

Gateway of last resort is not set

     115.0.0.0/8 is variably subnetted, 3 subnets, 2 masks
S       115.0.0.0/29 [1/0] via 192.168.0.2
S       115.0.11.0/24 [1/0] via 192.168.0.2
S       115.0.13.0/24 [1/0] via 192.168.0.2
     192.168.0.0/30 is subnetted, 1 subnets
C       192.168.0.0 is directly connected, FastEthernet0/24
C    192.168.10.0/24 is directly connected, Vlan10
C    192.168.20.0/24 is directly connected, Vlan20
```

Figure 3.29　Routing Table for MS0

It is also important to add two routes to the border router R0, so that R0 forwards data destined for the 115.0.11.0/24 and 115.0.13.0/24 networks to the ISP's routers. This is shown in Figure 3.30.

```
R0(config)#do sh ip rou
Codes: C - connected, S - static, I - IGRP, R - RIP, M - mobile, B - BGP
       D - EIGRP, EX - EIGRP external, O - OSPF, IA - OSPF inter area
       N1 - OSPF NSSA external type 1, N2 - OSPF NSSA external type 2
       E1 - OSPF external type 1, E2 - OSPF external type 2, E - EGP
       i - IS-IS, L1 - IS-IS level-1, L2 - IS-IS level-2, ia - IS-IS inter area
       * - candidate default, U - per-user static route, o - ODR
       P - periodic downloaded static route

Gateway of last resort is not set

     115.0.0.0/8 is variably subnetted, 3 subnets, 2 masks
C       115.0.0.0/29 is directly connected, FastEthernet1/0
S       115.0.11.0/24 [1/0] via 115.0.0.2
S       115.0.13.0/24 [1/0] via 115.0.0.2
     192.168.0.0/30 is subnetted, 1 subnets
C       192.168.0.0 is directly connected, FastEthernet0/0
S    192.168.10.0/24 [1/0] via 192.168.0.1
S    192.168.20.0/24 [1/0] via 192.168.0.1
```

Figure 3.30　Routing Table for R0

Check access to an extranet server from the intranet. We ping the web server on the extranet (115.0.13.10/24) from the Laptop3 host, and the result shows that the communication is normal, as shown in Figure 3.31.

```
C:\>ping 115.0.13.10

Pinging 115.0.13.10 with 32 bytes of data:

Reply from 115.0.13.10: bytes=32 time<1ms TTL=125
Reply from 115.0.13.10: bytes=32 time<1ms TTL=125
Reply from 115.0.13.10: bytes=32 time=11ms TTL=125
Reply from 115.0.13.10: bytes=32 time=15ms TTL=125
```

Figure 3.31　Result of Laptop3 Host Pinging the Web Server on the External Network

The question is, with millions of networks across the Internet, should all the millions of routing information be added to every Layer 3 device on the intranet? If we do have to do that, it would result in almost all of the Layer 3 devices on the intranet not being able to handle such a large routing table. The solution is clearly unacceptable to intelligent humans.

Route summarisation can solve this problem. We try to summarise the networks of the two routes into a larger network, and express the routing information in the routing table in terms of the summarised network, which can reduce the number of entries in the routing table and improve the efficiency of the query.

3. Route Summarisation: Summarises Route Entries from Multiple Networks into a Single

Step 1: Convert the Target Network of Route Entries to be Summarised Into A Binary Expression, as shown in Figure 3.32.

Target network number	Binary representation		Remark
115.0.11.0	01110011.00000000.00001	011.00000000	
115.0.13.0	01110011.00000000.00001	101.00000000	

Network bit part   Host bit part

Figure 3.32   Convert to Binary Expression

Step 2: Separate the Network Bit Part from the Host Bit Part to Get the Summarised Network.

Compare the binary number expressions of the two network addresses and separate the same part from the different part. The same part is both the network bit part of the aggregated network, and the different part is the host bit part of the aggregated network. By writing the host bit part as all "0", the network number of the aggregated network is obtained, as shown in Table 3.7.

Table 3.7   The Network Number of Aggregated Network

Network segment	Host Bit Sections	Decimal expression
01110011.00000000.00001	000.00000000	115.0.8.0/21

From this we can see that networks 115.0.11.0/24 and 115.0.13.0/24 are summarised to give a larger network with network number of 115.0.8.0/21. That is to say, in the routing table we replace the two original route records with the summarised routes by using the following command.

Step 3: Replace the Original Detail Routes with the Summarised Routes at the Layer 3 Device.

First delete the detail routes for networks 115.0.11.0/24 and 115.0.13.0/24 on MS0 and R0 with the following commands:

```
MS0(config)#no ip route 115.0.11.0 255.255.255.0 192.168.0.2
MS0(config)#no ip route 115.0.13.0 255.255.255.0 192.168.0.2
```

```
R0(config)#no ip route 115.0.11.0 255.255.255.0 115.0.0.2
R0(config)#no ip route 115.0.13.0 255.255.255.0 115.0.0.2
```

Configure the summarised route entries on MS0 and R0 with the following commands:

```
MS0(config)#ip route 115.0.8.0 255.255.248.0 192.168.0.2
```

It expresses that all data sent to the 115.0.8.0/21 network is forwarded to 192.168.0.2.

```
R0(config)#ip route 115.0.8.0 255.255.248.0 115.0.0.2
```

It expresses that data destined for network 115.0.8.0/21 is always forwarded to 115.0.0.2. Since networks 115.0.11.0/24 and 115.0.13.0/24 are included in network 115.0.8.0/21, the summarised routes can also forward data destined for networks 115.0.11.0/24 and 115.0.13.0/24.

The summarised routing table for core switch MS0 is shown in Figure 3.33.

```
MS0#sh ip rou
Codes: C - connected, S - static, I - IGRP, R - RIP, M - mobile, B - BGP
       D - EIGRP, EX - EIGRP external, O - OSPF, IA - OSPF inter area
       N1 - OSPF NSSA external type 1, N2 - OSPF NSSA external type 2
       E1 - OSPF external type 1, E2 - OSPF external type 2, E - EGP
       i - IS-IS, L1 - IS-IS level-1, L2 - IS-IS level-2, ia - IS-IS inter area
       * - candidate default, U - per-user static route, o - ODR
       P - periodic downloaded static route

Gateway of last resort is not set

     115.0.0.0/8 is variably subnetted, 2 subnets, 2 masks
S       115.0.0.0/29 [1/0] via 192.168.0.2
S       115.0.8.0/21 [1/0] via 192.168.0.2
     192.168.0.0/30 is subnetted, 1 subnets
C       192.168.0.0 is directly connected, FastEthernet0/24
C    192.168.10.0/24 is directly connected, Vlan10
C    192.168.20.0/24 is directly connected, Vlan20
```

Figure 3.33 Routing Table for Core Switch MS0

The routing table summarised by the border router R0 is shown in Figure 3.34.

```
R0#sh ip rou
Codes: C - connected, S - static, I - IGRP, R - RIP, M - mobile, B - BGP
       D - EIGRP, EX - EIGRP external, O - OSPF, IA - OSPF inter area
       N1 - OSPF NSSA external type 1, N2 - OSPF NSSA external type 2
       E1 - OSPF external type 1, E2 - OSPF external type 2, E - EGP
       i - IS-IS, L1 - IS-IS level-1, L2 - IS-IS level-2, ia - IS-IS inter area
       * - candidate default, U - per-user static route, o - ODR
       P - periodic downloaded static route

Gateway of last resort is not set

     115.0.0.0/8 is variably subnetted, 2 subnets, 2 masks
C       115.0.0.0/29 is directly connected, FastEthernet1/0
S       115.0.8.0/21 [1/0] via 115.0.0.2
     192.168.0.0/30 is subnetted, 1 subnets
C       192.168.0.0 is directly connected, FastEthernet0/0
S    192.168.10.0/24 [1/0] via 192.168.0.1
S    192.168.20.0/24 [1/0] via 192.168.0.1
```

Figure 3.34 Routing Table for Border Router R0

Here we are just summarising two routes into one. How should we summarise when there are millions of routes in the classroom Internet?

4. Re-summarise: Summarise Routing Entries for All Networks Into One

We can look for patterns in this aggregated mini-case and then make inferences. This is shown in Table 3.8.

Table 3.8 Internet Regional IP Address Information Table

Pre-aggregation		Aggregated	
Network number	Subnet mask	Network number	Subnet mask
115.0.11.0/24	/24	115.0.8.0	/21
115.0.13.0/24	/24		

As can be seen in Table 3.8, the length of the subnet mask before aggregation is/24 and the length of the network subnet mask after aggregation is/21-the length of the subnet mask has become shorter.

According to this law, the larger the network, the shorter the subnet mask should be, and when the network reaches its limit, the subnet mask should also be shortened to its limit. Therefore, all the networks of the world's organisations are aggregated to obtain the world's largest network (Internet), whose subnet mask must be the shortest (/0).

A subnet mask of /0 means that this network has no network bits but all host bits. The host bits of a network address are all "0s", so to express all the aggregated networks of the Internet, you can use 0.0.0.0/0. In other words, we can write a route to send all the data to the Internet.

First delete the summarised static routes on MS0 and R0:

Configure the summarised route entries on MS0 and R0 with the following commands:

```
MS0(config)#no ip route 115.0.8.0 255.255.248.0 192.168.0.2
R0(config)#no ip route 115.0.8.0 255.255.248.0 115.0.0.2
```

Configure the newly summarised routes on core switch MS0 with the following commands:

```
MS0(config)#ip route 0.0.0.0 0.0.0.0 192.168.0.2    //Forward all data to the 192.168.0.2 node
```

Configure the newly summarised routes on border router R0 with the following commands:

```
R0(config)#ip route 0.0.0.0 0.0.0.0 115.0.0.2    //Forward all data to node 115.0.0.2
```

The routing table on MS0 after the replacement is shown in Figure 3.35.

```
MS0#sh ip rou
Codes: C - connected, S - static, I - IGRP, R - RIP, M - mobile, B - BGP
       D - EIGRP, EX - EIGRP external, O - OSPF, IA - OSPF inter area
       N1 - OSPF NSSA external type 1, N2 - OSPF NSSA external type 2
       E1 - OSPF external type 1, E2 - OSPF external type 2, E - EGP
       i - IS-IS, L1 - IS-IS level-1, L2 - IS-IS level-2, ia - IS-IS inter area
       * - candidate default, U - per-user static route, o - ODR
       P - periodic downloaded static route

Gateway of last resort is 192.168.0.2 to network 0.0.0.0

     115.0.0.0/29 is subnetted, 1 subnets
S       115.0.0.0 [1/0] via 192.168.0.2
     192.168.0.0/30 is subnetted, 1 subnets
C       192.168.0.0 is directly connected, FastEthernet0/24
C    192.168.10.0/24 is directly connected, Vlan10
C    192.168.20.0/24 is directly connected, Vlan20
S*   0.0.0.0/0 [1/0] via 192.168.0.2
```

Figure 3.35   Routing Table on Core Switch MS0

The routing table on R0 is shown in Figure 3.36.

```
R0#sh ip rou
Codes: C - connected, S - static, I - IGRP, R - RIP, M - mobile, B - BGP
       D - EIGRP, EX - EIGRP external, O - OSPF, IA - OSPF inter area
       N1 - OSPF NSSA external type 1, N2 - OSPF NSSA external type 2
       E1 - OSPF external type 1, E2 - OSPF external type 2, E - EGP
       i - IS-IS, L1 - IS-IS level-1, L2 - IS-IS level-2, ia - IS-IS inter area
       * - candidate default, U - per-user static route, o - ODR
       P - periodic downloaded static route

Gateway of last resort is 115.0.0.2 to network 0.0.0.0

     115.0.0.0/29 is subnetted, 1 subnets
C       115.0.0.0 is directly connected, FastEthernet1/0
     192.168.0.0/30 is subnetted, 1 subnets
C       192.168.0.0 is directly connected, FastEthernet0/0
S    192.168.10.0/24 [1/0] via 192.168.0.1
S    192.168.20.0/24 [1/0] via 192.168.0.1
S*   0.0.0.0/0 [1/0] via 115.0.0.2
```

Figure 3.36   Routing Table for Border Router R0

The static route labelled 0.0.0.0/0 in the routing table is no longer labelled as "S", but as "S*", which is called the static default route.

Here, the static default route will be used to send all the data from the intranet to the Internet to the operator to send out, greatly streamlining the routing table of the intranet Layer 3 equipment, greatly improving the efficiency of the query routing table, but also reduces the management costs of the network. Another advantage is that if there are more networks on the Internet, the network administrator does not need to know which networks are added, what their network numbers are, and there is no need to add any additional routing entries to the Layer 3 devices on the intranet, because a default route can send all the data to the extranet.

However, it should be noted that default routes have a lower priority than detail routes. This is because the routers querging the routing table is based on the longest match principle for the subnet mask, and the default route has a subnet mask of 0. Logically, of course, it is the last to be matched. In other words, the router goes to the routing table to match the route entry based on the destination address on the packet, and if the detailed route entry cannot be matched, the default route will be the last one to be matched. At this point, the default route will forward the data according to the next hop IP address of the default route because it can be matched by all the destination networks, and will not result in the loss of the packet on this router.

Exercise: With reference to this task, delete all static routes in the intranet whose destination network is a public address and replace them with default routes. After the configuration is complete, users in the intranet can still access the ISP's router

Competence Expansion: Route summarisation simplifies the routing entries in the routing table and improves the efficiency of the router in matching routing entries, thus increasing the data forwarding rate. Simplicity is a kind of beauty, college students should be physically active, advocate a simple and moderate, green and low-carbon lifestyle, treat the ecological environment as we treat our lives, and make their own contributions to the construction of a beautiful China by leaving behind a living environment with blue sky, green earth and clear water.

Task 5: Deploying Domain Name Servers—Domain Name Service (DNS)

Deploying Domain Name Servers— Domain Name Service(DNS)

Task Objective: Understand the process of accessing a web server through a domain name.

After completing the configuration of tasks 1 to 3 of this project, we can use a browser in Laptop3 to access a Web server on an external network by typing the IP address of the Web server in the URL address bar, and the result is shown in Figure 3.37.

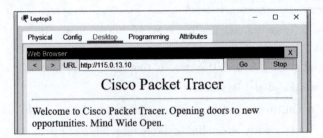

Figure 3.37 Laptop3 Accessing the Web Server Interface

How can we confirm that the web page accessed here is the one on the web server (115.0.13.10/24)? We can modify this web page on the web server, as shown in Figures 3.38 and 3.39.

Change "Cisco Packet Tracer" to "baidu" and "baidu.com" in two places in the original webpage, and click "Save" in the lower right corner to save. When prompted whether to overwrite, click "Yes" to confirm the overwrite. Browse the web server on Laptop3 again, and the page you see will be changed, as shown in Figure 3.40.

Project 3   Access the Internet

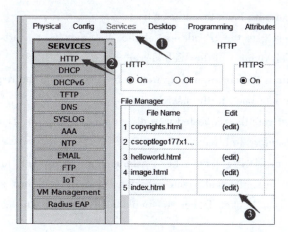

Figure 3.38   Modifying the index. html file on the Web server

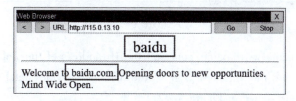

Figure 3.39   Modifying the Default Home Page Information

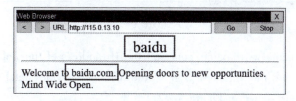

Figure 3.40   Browsing the Web Server Again on Laptop3

The web page displayed has changed, but the change is the result of a previous modification. Then you can be sure that the web page seen on Laptop3 is that of the web server (115.0.13.10/24).

Normally, we do not use the IP address to access the server, because we cannot remember the IP address of the server due to its irregularity. In general, we access the Web host by its name-the domain name. Just like classmates generally call each other by names but not by school numbers, because the school number is not easy to remember.

Users access the Web host through the domain name of the process is as follows:

①The user enters the domain name of the Web server through the address bar, and the host acquires the domain name that the user wants to access.

②The user host sends the acquired domain name to the DNS server and requests the DNS server to query the IP address corresponding to the domain name. Because the host encapsulates the packet when the destination address is the IP address of the Web server, not its domain name.

③After the DNS server finds the IP address corresponding to the domain name of the Web server, it sends the IP address back to the user host.

④After the user host gets the IP address of the Web server, it starts to encapsulate the IP packet and sends the Web request to the Web server.

⑤The web server finds the corresponding web page according to the request from the user host and sends the web page data back to the user host.

⑥The end-user host gets the requested web page data and displays it in the browser. The user can then see the corresponding web page.

Domain Name System (DNS) servers keep a table that corresponds to the domain name of a Web host and its IP address. Normally you can use Windows or Linux servers to provide DNS service, here we use Cisco Packet Tracer to simulate DNS server, its configuration interface is shown in Figure 3.41.

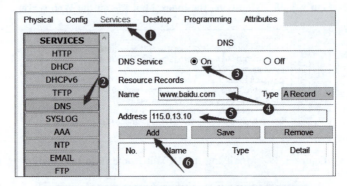

Figure 3.41 DNS Service Configuration interface

Enter the domain name www.baidu.com and the IP address of the web server, click "Add" button to add records to the database, after the completion the database will increase the corresponding records, as shown in Figure 3.42.

Figure 3.42 Mapping Table of DNS Server Domain Names and IP Addresses

After adding successfully, you also need to fill in the IP address of the DNS server in the location where the user sets the parameters of the NIC. Tell the user host to send the domain name to 115.0.11.10/24 if there is a domain name that needs to be resolved to an IP address, as shown in Figure 3.43.

After completing all the configurations, users can access the web server by typing the domain name (www.baidu.com) in the address bar of the browser on the Laptop3, and the result is shown in Figure 3.44.

170

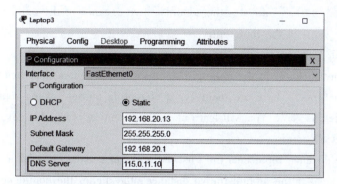

Figure 3.43   Configuring the IP Address of the DNS Server

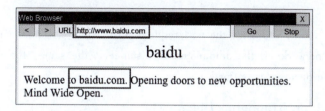

Figure 3.44   User Access to Web Server Results Via Domain Name

The results show that users can access the Web server with a domain name with the help of DNS service.

Exercise: With reference to this task, deploy Web and DNS servers on the Internet, configure them appropriately, and realise that users on the intranet can access the Web servers with domain names with the help of DNS services.

Competence Expansion: In the Internet, we use the IP address and domain name to identify a server, but since domain name is more convenient to remember, we use DNS servers to help us to resolve the domain name into IP address so that IP packets can be encapsulated successfully. But with users relying so heavily on DNS services, if a DNS server is hijacked, it will cause a large number of network services to come to a standstill. This kind of damage to network communication security is illegal, college students should carry forward the spirit of the socialist rule of law, the construction of the socialist culture of the rule of law, to enhance the whole society to practice the rule of law of the positive and proactive, the formation of law-abiding is honourable, the social atmosphere of shameful law-breaking.

 ## Summary and Expansion

1. Interpretation of Static Routing Information

```
S 115.0.0.0 [1/0] via 192.168.0.2
```

①"S" is short for Static, which means Static Route.

②"115.0.0.0" indicates the network number of the target network.

③The "1" in "[1/0]" indicates the administrative distance of the route entry, and the "0" indicates the metric value of the route entry.

④"192.168.0.2" indicates the IP address of the next hop node.

The routing entries stored in the routing table are what the router considers to be the best path to the destination network. There are many types of routing entries, static routes are just one of them and are labelled as "S". You will learn more about routing protocols later, such as R for RIP routing and O for OSPF routing.

Administrative distance represents the trustworthiness of a route entry. For routes reaching the same network, the smaller the administrative distance, the higher the confidence level, indicating that the route entry is superior. And only routes that are considered optimal can be entered into the routing table.

The metric of a route, indicates the cost required to reach a destination network. For routes that reach the same network, if the administrative distance is the same, the route entry with the smaller administrative distance is optimal.

2. An Analogy on Managing Distance

There is a good example to help us understand the sentence "For routes reaching the same network, the smaller the administrative distance, the higher the trustworthiness": someone is lost in an unfamiliar city, so he asks for directions at an intersection and how to go to a certain place He asks several persons consecutively, and they all point to different directions, and at this point, who should he trust? Who should the lost person believe in? The answer is: whoever has the highest credibility.

Trustworthiness should also have an indicator to measure, this indicator is called "management distance".

If the administrator configures several entries to reach the same network, and the next hop is not the same, how can the router choose the best route among these several paths to reach the same network? The router considers the route entry with the smallest administrative distance to be the most reliable, and that entry is considered to be the best route and selected into the routing table. If there is more than one route with the smallest administrative distance, then their metrics are compared, and the one with the smallest metric is considered the best route and is selected and placed into the routing table.

3. The Longest-match Principle and the Position of the Default Route in the Routing Table

The way in which a router matches route entries is called the "longest match principle". When a router receives a packet, it reads the destination address on the packet and matches the network number in the routing table. If it finds multiple entries with the same network number, it will match the one with the longest subnet mask.

According to this rule, for a default route with a subnet mask of 0, it will always the last to be matched, because 0 is the shortest subnet mask.

4. Modify the Administrative Distance of Static Routes

The default administrative distance for static routes is 0, but it is possible to change if needed. The following is the command to change the administrative distance of a static route to the 115.0.11.0/24 network to 100:

```
MS0(config)# ip route 115.0.11.0 255.255.255.0 192.168.0.2 100
```

The routing table is examined and the results are shown in Figure 3.45.

```
Gateway of last resort is 192.168.0.2 to network 0.0.0.0

     115.0.0.0/8 is variably subnetted, 2 subnets, 2 masks
S       115.0.0.0/29 [1/0] via 192.168.0.2
S       115.0.11.0/24 [100/0] via 192.168.0.2
     192.168.0.0/30 is subnetted, 1 subnets
C       192.168.0.0 is directly connected, FastEthernet0/24
C    192.168.10.0/24 is directly connected, Vlan10
C    192.168.20.0/24 is directly connected, Vlan20
S*   0.0.0.0/0 [1/0] via 192.168.0.2
```

Figure 3.45 Routing Table for MS0

As you can see in the routing table, the administrative distance of the route entry to the 115.0.11.0/24 network has been changed to 100. Generally, in a multi-exit network, we will configure multiple static routes with different administrative distances. Normally the route with the smallest administrative distance is enabled, and when that route fails, the route with the second smallest administrative distance is enabled, thus achieving the path backup, which is called the floating static route technique.

5. Configuring Static Routes with Outgoing Ports

There are two ways to configure static routes: static routes with hops and static routes with outgoing ports. In Task 1 of this project, we adopt the first way, which is also the most commonly-used way, as long as the address of the next hop is reachable, the route will be available, the disadvantage is that it needs to be recursively queried, which occupies more system resources.

The following is the way to configure a static route by using a static route with a forwarding port. According to the topology shown in Figure 3.1, if MS0 has data to send to 115.0.0.0/29, the data should be sent from port Fa0/24. So it can be configured like this:

```
MS0(config)#ip route 115.0.0.0 255.255.255.248 fastEthernet 0/24
```

The routing table is examined and the results are shown in Figure 3.46 below.

The "S 115.0.0.0/29 is directly connected, Fast Ethernet0/24" routing entry tells switch MS0 that if there is data to be sent to the 115.0.0.0/29 network, it can be forwarded out of the FastEthernet 0/24 port, the end result is the same. The advantage is that the router knows directly from the route entry to forward the data out of port FastEthernet 0/24 without recursive queries,

```
MS0(config)#do sh ip rou
Codes: C - connected, S - static, I - IGRP, R - RIP, M - mobile, B - BGP
       D - EIGRP, EX - EIGRP external, O - OSPF, IA - OSPF inter area
       N1 - OSPF NSSA external type 1, N2 - OSPF NSSA external type 2
       E1 - OSPF external type 1, E2 - OSPF external type 2, E - EGP
       i - IS-IS, L1 - IS-IS level-1, L2 - IS-IS level-2, ia - IS-IS inter area
       * - candidate default, U - per-user static route, o - ODR
       P - periodic downloaded static route

Gateway of last resort is 192.168.0.2 to network 0.0.0.0

     115.0.0.0/8 is variably subnetted, 2 subnets, 2 masks
S       115.0.0.0/29 is directly connected, FastEthernet0/24
S       115.0.11.0/24 [100/0] via 192.168.0.2
     192.168.0.0/30 is subnetted, 1 subnets
C       192.168.0.0 is directly connected, FastEthernet0/24
C    192.168.10.0/24 is directly connected, Vlan10
C    192.168.20.0/24 is directly connected, Vlan20
S*   0.0.0.0/0 [1/0] via 192.168.0.2
```

Figure 3.46　Routing Table for Core Switch MS0

which improves the data forwarding efficiency.

There are two drawbacks to configuring static routes with the outgoing port. One is that when the forwarding port fails, the path will fail even if the next hop address is reachable. The second is that although it is labelled as S (static route) here, the 115.0.0.0/29 network is shown later as a directly-connected network, which is obviously an error. It will make switch MS0 mistakenly think that network 115.0.0.0/29 is directly connected to it, and it will send an ARP request asking for the MAC address of the destination host. But in fact the destination host is not directly connected but needs to span several networks.

The ARP request is sent out as a broadcast, and if the forwarding port is a broadcast-type network, it will lead to the excessive traffic on the network which may exhaust the router's memory, thus affecting normal network communication. The solution to the problem is to specify both the forwarding port and the address of the next hop with the command of ip route 115.0.0.0 255.255.255.248 FastEthernet0/24 192.168.0.2 (this command is not supported by the Cisco Packet Tracer emulator).

6. Several Types of NAT

There are three types of NAT: static NAT, dynamic NAT, and PAT.

This project uses two types of NAT: static NAT and PAT, which are also the most commonly-used NAT techniques. Static NAT is used to map the IP address of the internal server to a public IP address, so that users from outside the network can access the internal server. PAT technology is used to enable the internal user hosts to access the resources of the outside network, which achieves the reuse of the public IP address through the way of port mapping and saves the public IP address.

Another way to implement dynamic NAT is to create a public IP address pool and map the IP address of an internal user to a free public address in the pool. The command is as follows:

R0(config)# ip nat pool NAT_POOL 115.0.0.3 115.0.0.6 netmask 255.255.255.248   //Define an address pool with the name NAT_POOL (Arbitrary value)

R0(config)# ip nat inside source list 1 pool NAT_POOL   //Map user IPs defined by the list to address pools (note that the list serial number and address pool name must match exactly as defined)

If there is no free public IP address in the address pool, the user data can only wait, thus affecting the efficiency of data communication. The solution to this problem can only be paying attention to the observation of the use of the address pool to avoid depletion, or using the overload parameter to allow overload conversion. The command is as follows:

R0(config)# ip nat inside source list 1 pool NAT_POOL overload   //Allow NAT overload

7. Domain Name Services

The domain name service is implemented by using the DNS protocol, which uses a hierarchical system to create a database to provide the name resolution, as shown in Figure 3.47. DNS uses domain names to divide the hierarchy.

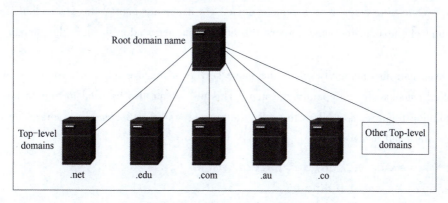

Figure 3.47   Domain Name Structure

The domain name structure is divided into smaller managed domains. Each DNS server maintains specific database files and is responsible for managing the "domain-IP" mapping for only that small portion of the DNS structure. When a DNS server receives a request to convert a domain name that does not belong to the DNS zone for which it is responsible, the DNS server can forward the request to a DNS server in the zone that corresponds to the request for conversion. DNS is scalable because the host name resolution is spread across multiple servers.

Different top-level domains have different meanings, representing the types of organisations or orgins of countries/regions. See Table 3.9 for an example.

Table 3.9 Meaning of Common Domain Name Suffixes

Serial number	Top-level domains	Hidden meaning
1	.net	.net is one of the earliest and most popular generic domain names used internationally
2	.edu	edu is a short form of education. It generally indicates an educational institution, such as a university
3	.com	com is short for company, which means company enterprise. It is one of the most widely popular generic domain name suffixes in the world
4	.org	Applicable to all types of organisations, including not-for-profit groups
5	.cn	The ".cn" domain name is an international top-level domain name managed by China, and is China's own Internet identity. cn represents China, which embodies a cultural identity, its own value and positioning
6	.gov.cn	gov is the abbreviation of government. The ".gov.cn" domain name is an important identifier for the websites of Chinese government agencies and other government departments, and is specifically used for government agencies and other departments in China

We usually configure network devices with one or more DNS server addresses that a DNS client can use for domain name resolution. ISPs often provide addresses for DNS servers. When a user application requests to connect to a remote device via a domain name, the DNS client requests a query to one of the domain name servers to obtain a numeric address after the domain name is resolved.

The user can also manually query the name servers to resolve a given host name by using a utility called nslookup in the operating system. This utility can also be used to troubleshoot domain name resolution failures, as well as to verify the current status of the name servers. The reference commands are listed below:

```
C:\Users\Administrator>nslookup
Default server: UnKnown
Address: fe80::1
> www.baidu.com
Server: UnKnown
Address: fe80::1
Non-authoritative responses.:
Non-authoritative responses: www.a.shifen.com
Addresses: 183.232.231.174
          183.232.231.172
Aliases: www.baidu.com
>
```

The results of the DNS query show the domain name www.baidu.com The mapped IP addresses are 183. 232. 231. 174 and 183. 232. 231. 172.

 Thinking and Training

Continue to complete the following requirements based on Question 7 in Reflection and Training of the Project 2 as per the topology shown in Figure 3.48:

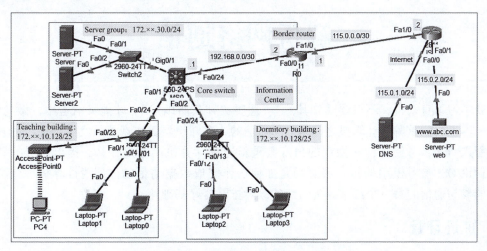

Figure 3. 48 Project Topology

(1) Plan the IP addresses of each department in the intranet and configure them appropriately to enable users in both the teaching building and dormitory building to access the university's servers.

(2) Configure static routes in the campus intranet to achieve campus-wide interoperability. (Hint: All users should be able to access the intranet ports of the border routers.)

(3) Configure the Layer 3 devices on the intranet with default routes to send packets destined for the Internet to the ISP.

(4) Apply PAT technology in the border router R0 to achieve that all users in the intranet can access the servers of the Internet.

(5) Apply static NAT technology in border router R0 to enable external servers to access the Web interfaces of the two internal servers through browsers.

(6) Add the record of www. abc. com to the DNS server of the Internet to ensure that users of the internal network can access the external Web server through the browser, using the domain name.

Note: There is no permission to configure any device on the Internet, including the ISP's router.

项目 4

解析网络通信

就学习本身来说,不仅要"知其然",还要"知其所以然",才能灵活运用,做到触类旁通。本项目侧重解释计算机之间是如何通信的,在知识层面实现了从实践到理论的升华。通过引入 OSI 参考模型,从微观的角度去重新诠释计算机网络是什么。提出分层的概念,详细讨论 OSI 参考模型的每一层之间是如何协作完成数据通信的,理清计算机、交换机、路由器参与通信的每一个细节,为后续的故障排除打好基础。

知识目标

(1) 了解常用的网络模型(OSI 参考模型和 TCP/IP 网络模型);
(2) 理解数据通信过程中 OSI 参考模型中各层的作用以及各层之间的协作;
(3) 理解交换机和路由器在通信过程中的作用;
(4) 理解可变长子网掩码(Variable Length Subnet Mask,VLSM)的意义。

技能目标

(1) 能描述 OSI 参考模型各层的功能;
(2) 能基于 OSI 参考模型清晰描述完整的数据通信过程;
(3) 掌握 VLSM 方法,将大网络分割成小网络。

相关知识

通过项目 1 的学习,我们学习了计算机网络是什么。但其实那是从宏观的、具体的角度去描述计算机网络,那样更容易让初学的人去理解。但是如果要弄清楚网络是如何工作的,数据是如何从发送方一步一步转发到接收方的,整个通信的每个工作细节是怎样的,我们就要从另外一个角度去描述什么是计算机网络。跟宏观的、具体的角度不一样,这个角度应该更加抽象和专业化,但它确实能帮助我们更好地理解数据通信过程。

有人可能会疑惑:我们为何要去关心数据通信的细节呢?只要网络设备能帮用户把数据送到接收方就可以了,数据传输任务应该完全由网络中间设备去承担!这里需要理清我们未来的身份——网络工程师,而不是普通的网络用户。当网络出现各种故障的时候,网络工程师就要在最短的时间排除故障,恢复网络通信,这就像医生的职责就是治病救人一样。但是如果这个医生对人体的各器官工作细节以及器官之间如何协同工作以保障伟大的生命工程都不清楚，相信几乎没有人敢找这样的医生看病。同样的，如果我们不理解网络工作的原理，不理解数据通信的细节，当网络出现故障时，我们就无法判断故障点在哪里，是什么原因导致了这个故障，该如何解决，今后该如何避免。会有客户找这种不理解通信原理的人来帮助解决网络通信问题吗？几乎不会。

为了更专业地解释什么是计算机网络，我们引入了网络模型的概念。

网络模型只是计算机网络工作原理的表示方法而并非实际的网络。其从网络功能的角度，描述特定的层需要完成什么，但不规定如何完成，以此来保持各类网络协议和服务中的开放性和一致性。

网络模型有很多，比较常用的有两个：OSI 参考模型和 TCP/IP 网络模型。

任务 1：初识 OSI 参考模型

任务目标：了解 OSI 参考模型的分层思路以及各层的功能。

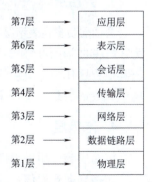
初识 OSI 网络参考模型

计算机网络刚面世的时候，通常只有同一个厂家生产的计算机才能相互通信。一个企业部署一个网络的时候，必须全部采购同一家公司的产品，不能集成几个厂家的设备。直到 20 世纪 70 年代末期，国际标准化组织（International Organization for Standardization，ISO）开发了开放式系统互连（Open System Interconnect，OSI）参考模型。该模型目的是以协议的形式帮助厂商生产可互操作的网络设备和软件，让不同厂家的网络能相互通信。

OSI 参考模型是主要的网络架构模型，描述了数据和网络信息如何通过网络介质从一台计算机的应用程序传输到另一台计算机的应用程序。为了清楚地描述数据传输的细节，OSI 参考模型把一次网络通信过程的所有工作任务进行统计并分组（分层），规定了每层都需要完成哪些工作进行层和层之间相互协作以共同完成整个通信任务。就像物流公司为了完成快递的传输，成立了好几个部门，规定了每个部门应该完成的任务，各部门各司其职又相互协作，最终完成一个快递的传输。这里 OSI 参考模型定义的层，就类似于快递公司成立的业务部门。

OSI 参考模型定义了 7 个层：物理层、数据链路层、网络层、传输层、会话层、表示层、应用层，如图 4.1 所示。

OSI 参考模型的 7 层被分成两组，第 1 组：物理层、数据链路层、网络层；第 2 组：传输层、会话层、表示层、应用层。第 1 组负责的是端到端的数据传输，第 2 组负责应用程序之间的通信。也就是说，第 1 组负责数据从这台主机发往另一台主机，至于数据是主机中哪个应用程序发出的，对端主机哪个应用程序接收，它是不关心的。这个类似于快递公司负责货物中转的大货车司机，他不会去关心车厢内的快递是从哪里来的，要发给谁，他负责的只是像接力一样把货物从这个城市发到那个城市；第 2 组负责运行在终端主机上的应用程序之间的相互通信，同时还要提供与终端用户的交互，这个工作类似快递公司负责接收快递的员工所要完成的任务。与开大货车的司机不一样，快递员要

图 4.1 OSI 参考模型

确定发出快递用户的身份，负责派件的员工要在区域内负责找到接收快递的人，凭有效证件领取快递。

参考模型每一层所要完成的工作是不一样的，各层之间又要相互协同，共同完成一次数据传输。就像快递公司不同岗位的员工所负责的工作都不一样，各岗位人员共同协作，任何一个角色缺一不可，一起完成快递的传输。

参考模型每一层都需要完成什么工作呢？OSI 参考模型各层的功能如下。

1. 应用层

应用层是用户与计算机交互的通道，当需要访问网络的时候，这一层就开始发挥作用。工作在应用层的协议包括 HTTP、DNS、DHCP、FTP、SMTP 和 POP3 等。

但有些应用程序并不是工作在应用层，比如 IE 浏览器。即使把网络协议和网卡全部禁用，IE 也可以浏览本地文件，但是一旦它收到用户请求需要访问网络资源的时候，它才去调用应用层的端口和服务，此时，服务才是应用层的程序。

应用层还有一个很重要的任务，就是要负责目标通信方的可用性，并判断是否有足够的资源确保想要的通信。

2. 表示层

OSI 在表示层定义了如何格式化标准数据，以便能确保从一个系统的应用层传来的数据可以被另外一个系统所识别和读取。

表示层为应用层提供服务。在发送方，表示层将应用层传下来的数据进行数据转换和代码格式化，然后将处理后的数据发给下一层（会话层）。对于接收方，表示层就要负责将会话层发来的格式化数据进行逆向处理，还原为最原始的数据，然后转发给应用层。

数据加密、解密、压缩和解压缩等都和表示层有关。发送方的表示层根据需要，将应用层传下来的数据进行加密和压缩；接收方的表示层从下层（会话层）得到的数据如果是经过压缩的，则要做解压缩的工作，如果是加密的数据，则要对数据进行解密，最后将处理完的数据发送给应用层。

3. 会话层

会话层负责在表示层实体之间建立、管理和终止会话，它协调和组织系统之间的通信。其为客户端的应用程序提供了打开、关闭和管理会话的机制，即半永久的对话。会话的实体包含了对其他程序作会话链接的要求及回应其他程序提出的会话链接要求。在应用程序的运行环境中，会话层是这些程序用来提出远程过程调用（Remote Procedure Calls，RPC）的地方。

4. 传输层

传输层为应用进程提供端到端的通信服务，可以包括以下功能。

连接导向式通信：通常对于一个应用进程来说，把连接解读为数据流而非处理底层的无连接模型更加容易。

相同次序交付：网络层通常不保证数据包到达顺序与发送顺序相同。这通常是通过给报文段编号来完成的，接收者按次序将它们传给应用进程。

可靠性：由于网络拥塞和错误，数据包可能在传输过程中丢失。通过检错码（如校验和），传输协议可以检查数据是否损坏，并通过向发送者传 ACK 或 NACK 消息确认是否正确接收。自动重发请求方案可用于重新传输丢失或损坏的数据。

流量控制：有时必须控制两个节点之间的数据传输速率以阻止快速的发送者传输超出接收缓冲器所能承受的数据，造成缓冲区溢出。这也可以通过减少缓冲区不足来提高效率。

拥塞避免：拥塞控制可以控制进入到电信网络的流量。

多路复用：端口可以在单个节点上提供多个端点。例如，邮政地址的名称是一种多路复用，并区分同一位置的不同收件人。每个计算机应用进程会监听它们自己的端口，这使得在同一时间可以使用多个网络服务。

为了完成传输层的工作，人们设计了 TCP 和 UDP 两个协议。这两个协议的工作是并行的，是不相干的。而且它们完成传输层所定义的工作不一样：TCP 能通过各种策略确保数据的可靠传输，但是 UDP 忽略了很多细节，通过牺牲可靠性，提高了数据传输效率。程序设计人员通常要根据所要传输数据的重要性决定要选用 TCP 还是 UDP 负责传输。

5. 网络层

网络层位于数据链路层与传输层之间，为传输层提供服务。它将传输层发给它的数据段进行封装，生成 IP 数据包；同时将数据链路层发上来的 IP 数据包进行解封装，将数据段往上发给传输层。

网络层主要提供路由和寻址的功能，使两终端系统能够互连且决定最佳路径，并具有一定的拥塞控制和流量控制的能力，相当于发送邮件时需要地址一般重要。由于 TCP/IP 协议体系中的网络层功能由 IP 规定和实现，故又称 IP 层。

网络层具备两个功能：

寻址：网络层使用 IP 地址来唯一标识互联网上的设备，其依靠 IP 地址实现主机间的相互通信。

路由：不同的网络之间相互通信必须借助路由器等三层设备，网络层在三层设备创建一个路由表，在路由表中保存到达某个网络的最佳路径。当有数据包需要传输时，网络层就必须根据路由表做出数据转发的决定。

工作在网络层的网络设备主要有：路由器、三层交换机。

6. 数据链路层

数据链路层在两个网络实体之间提供数据链路连接的创建、维持和释放管理。封装数据链路数据单元——帧，并对帧定界、同步、收发顺序的控制。数据链路层同时还负责传输过程中的流量控制、差错检测和差错控制等方面。

数据链路层位于物理层与网络层之间，它为网络层提供服务，会将网络层发下来的 IP 数据包封装成数据帧，帧的格式由物理层的传输介质决定。对于不同的传输介质，数据链路层封装的帧格式是不一样的。

为了确保数据传输的安全性，它在帧的尾部放置校验码（Parity，Sum，CRC）以检查帧在传输过程中是否被破坏，从而将物理层提供的可能出错的物理连接改造成逻辑上无差错的数据链路。

以太网的数据链路层还负责控制对介质的访问。通过载波侦听多路访问/冲突检测（Carrier Sense Multiple Access with Collision Detection，CSMA/CD）和载波侦听多路访问/冲突避免（Carrier Sense Multiple Access with Collision Avoidance，CSMA/CA）机制决定如何去访问冲突域的传输介质，以提高数据传输效率。数据链路层只负责传输介质的一端到另一端的数据传输。

工作在链路层的设备主要有：网桥、交换机。

7. 物理层

物理层是计算机网络 OSI 参考模型中最底层，也是最基础的一层。简单地说，网络的物理层负责在各种物理媒体上发送和接收数据比特（bit，数据位）。

物理层定义了要在终端系统之间激活、维护和断开物理链路而需要满足的电气、机械、规程和功能的需求。它以标准的形式规定了物理层接头和各种物理拓扑，让不同的系统能够彼此通信。比如网络端口，如果不统一标准，不同厂家生产的设备端口和接头大小不一，或者每根引脚功能不同，显然无法兼容。

工作在物理层的设备主要有集线器，但是由于其工作效率低，已经被淘汰，在此不做讨论。

课堂练习：用自己的语言简单描述 OSI 参考模型各层的功能。

素质拓展：OSI 参考模型将网络数据传输的整个任务过程分成了 7 块，每一块工作分别由一层完成，一共 7 层，每一层各司其职。通信过程中，每一层都是不可或缺的，但是某一层的工作如果和整个模型割裂开来，是没有任何意义的。就像在社会中，人既作为个体而存在，又作为集体中的一员而存在，集体和个人是不能分割的。国家利益、社会整体利益和个人利益是辩证统一的。

任务 2：解析一次简单的数据通信过程——OSI 各层的协作

解析一次简单的数据通信过程——OSI各层的协作

任务目标：理解 OSI 参考模型各层之间相互协作完成一次基础的数据通信过程。

OSI 各层之间如何协同完成数据通信呢？我们通过最简单也是最常体验到的例子进行描述——浏览网页的过程：

①用户打开 IE 浏览器，在地址栏输入服务器的地址。

②IE 浏览器调用应用层的 HTTP 生成一个请求包，准备发给服务器，请求相关的网页。

③服务器收到客户端发来的请求后，去搜索客户请求的网页，最后将网页数据发回给客户端。

④客户端拿到网页数据后，在 IE 浏览器展示出来，用户就可以看到请求的网页。

如图 4.2 所示是客户端将 HTTP 请求发给 Web 服务器的整个过程。

如表 4.1 所示是数据通信过程，其基于 OSI 参考模型，描述客户端将 HTTP 请求发给 Web 服务器的过程。

项目 4 解析网络通信

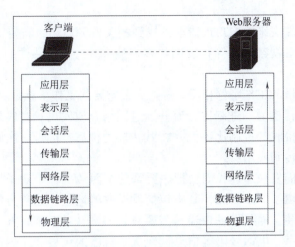

图 4.2 客户端将 HTTP 请求发给 Web 服务器的整个过程

表 4.1 数据通信过程

参考模型	客户端	Web 服务器
应用层	（1）生成 HTTP 请求数据，将数据往下发，给表示层	（14）HTTP 服务拿到请求数据后，根据客户的请求查找相关的资源
表示层	（2）进行数据格式化，根据需求完成压缩、加密，将处理完的数据往下传，给会话层	（13）表示层拿到数据后，如果数据是压缩的，则需要解压缩；如果数据是加密的，则需要进行解密，得到最原始的数据——HTTP 请求数据。最后将数据往上发给应用层的 HTTP 服务
会话层	（3）启动一个会话，将数据往下传给传输层	（12）拿到数据后，将数据往上发给表示层
传输层	（4）将数据交给 TCP 处理。由 TCP 负责对数据进行分片，并对每个数据片进行编号；接着将数据片封装成数据段，在数据段的头部填写各种参数，包括源端口号和目的端口号；三次握手完成后，将数据段往下传给网络层	（11）拿到数据段后，将数据段解封装，得到数据片，然后给发送方的 TCP 返回一个确认信息，表示已经收到数据；同时根据数据段头部的序号进行重组，得到最初完整的数据，并根据目的端口号将数据往上发送，给会话层
网络层	（5）网络层拿到数据段后，将数据段封装成 IP 数据包。在 IP 数据包的头部填写各种参数，包括源 IP 地址和目的 IP 地址，最后将数据包往下传给数据链路层	（10）拿到 IP 数据包后，查看 IP 数据包头部的目的 IP 地址。如果无权接受，则把 IP 包丢弃；如果有权接受，则对 IP 数据包进行解封装，得到数据段，将数据段往上发给传输层
数据链路层	（6）数据链路层拿到 IP 数据包后，将 IP 数据包封装成数据帧。在数据帧的头部和尾部填写各种参数，其中包括源 MAC 地址和目的 MAC 地址，最后将数据帧往下传，给物理层	（9）从物理层拿到数据帧，查看数据帧头部的目的 MAC 地址，如果无权接受，则将数据帧丢弃；如果数据帧头部的目的 MAC 地址是自己的或者是广播 MAC 地址，则接受，并将数据帧解封装，得到 IP 数据包，完成校验后，将 IP 数据包往上传，发给网络层
物理层	（7）物理层将数据帧转换为二进制比特流，通过介质发送出去	（8）接收到二进制比特流后，将二进制比特流转换为数据帧，将数据帧往上发送给上一层

183

服务器根据客户端发来的 HTTP 请求，找到相关资源以后，将资源数据（可能是一个网页，一张图片或者音视频媒体等）发回给客户端。其过程是一样的，只是方向刚好和请求包发送的方向相反。

发送方和接收方做的工作刚好相反。发送方最重要的工作，就是将数据进行三次封装，最后在物理层将数据转换为二进制比特流并通过网络介质发出；对应地，接收方最重要的工作就是从网络介质接收二进制比特流并将其转换为帧后，进行三次解封装，最后得到发送方生成的原始数据传到应用层。

这个过程就像我们发快递。我们会把想要寄的东西装到一个盒子里，然后给这个盒子贴上一张单子，上面填了好多信息，其中包括发送方的地址和接收方的地址。区别是，网络通信要进行三次封装，而我们发快递只需要封装一次。对于接收方，在快递的单子上看到是自己的地址、名字、电话号码等信息，确认是自己的包裹，才会拆开。网络通信中的接收方也要做相应的工作，确认是自己的数据以后，对应地在数据链路层、网络层和传输层一共进行三次解封装，拿到最原始的数据。

如果一定要拿网络数据的转发和快递的传送做一个类比，那么按网络数据传输的方式传送快递，就是把要寄送的货物先封装在小盒子，贴个单子，然后再将小盒子放入中盒子，贴第二个单子，最后将中盒子放入大盒子，贴第三张单子，完成三次封装的过程。接收方拿到快递，则要先拆开大盒子，再拆开中盒子，最后拆开小盒子，最后拿到发送方发送的货物。

为什么要封装这么多次？简单地说，为了确保数据能顺利传输。比如有人从国外网购一部手机，商家并不是将手机直接丢到船上，经历长时间的风吹雨打日晒送到用户手中的。而是为了保护好手机，厂家先将手机封装到手机盒子里，然后商家将手机盒子装到快递盒子里，最后上船的时候，还要将快递盒子装入集装箱。每一次封装都是有价值和意义的，是经过精心设计的，不是想当然的或可有可无的。参考模型每一层处理数据的方式不一样，需要不同的封装。即使在同一层，不同的协议封装的数据单元也是不一样的。

OSI 参考模型三次封装生成的协议数据单元（Protocol Data Unit，PDU）的名称都是不一样的。第一次封装由传输层完成，封装后的 PDU 叫数据段（Segment）；第二次封装由网络层完成，封装后的 PDU 叫数据包（Package）；第三次封装由数据链路层完成，封装后的 PDU 叫数据帧（Frame）。

课堂练习：用自己的语言描述一次完整的数据通信过程。

素质拓展：OSI 参考模型一共 7 层，每一层各司其职，同时各层与上、下层协同，完成每次数据通信。我们除了专注干好本职工作，还要深刻地意识到，各行业的分工越来越细，几乎每项工作或任务都需要团队相互协作来完成。我们要正确认识团队和个人的辩证统一关系：一方面，个人离不开团队，团队把每个劳动者的智慧和力量凝聚在一起，形成巨大的创造力；另一方面，团队是由若干个人组成的，不调动个人的积极性，也就不会有集体的创造力。

任务 3：交换机参与数据通信——数据链路层

任务目标：理解交换机的工作过程。

交换机工作在 OSI 参考模型的数据链路层，负责处理数据帧，是计算机网络中使用最多的设备。因为它负责的是 OSI 参考模型第二层的工作，所以有时候习惯上被称为二层设备。

交换机参与数据通信——数据链路层

1. 交换机依靠 MAC 地址表转发数据帧

交换机负责的工作很简单，就是将数据帧从某个端口接收进来，然后从另一个端口转发出去。问题是：交换机有这么多个端口，其如何决定要从哪个端口将数据帧转发出去？

如图 4.3 所示为研发部的网络拓扑图。Laptop2 和 Laptop3 主机的 MAC 地址分别为 0001.42d1.e839 和 0060.5c74.2c99。

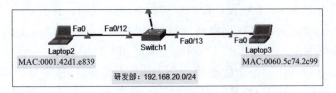

图 4.3 研发部的网络拓扑图

交换机的缓存中有一个记录主机 MAC 地址信息的表（Mac Address Table），在交换机的特权模式下可以用 show mac-address-table 命令查看，如图 4.4 所示。

```
Switch1#show mac-address-table
        Mac Address Table
-------------------------------------------
Vlan    Mac Address       Type        Ports
----    -----------       --------    -----
 1      0001.c774.3002    DYNAMIC     Fa0/24
 20     0001.42d1.e839    DYNAMIC     Fa0/12
 20     0060.5c74.2c99    DYNAMIC     Fa0/13
```

图 4.4 交换机中的 MAC 地址表

MAC 地址表中各字段的意义如表 4.2 所示。

表 4.2 MAC 地址表中各字段的意义

序号	字段	意义	备注
1	Vlan	端口所在的 VLAN	
2	Mac Address	接入该端口的主机的 MAC 地址	
3	Type	获得该主机 MAC 地址的方式（类型）	DYNAMIC：动态
4	Ports	端口号	

如果 Laptop2 要发送一个数据帧给 Laptop3，过程如下：

①Laptop2 在封装数据帧的时候，源 MAC 地址为自己的 MAC 地址，目的 MAC 地址为 Laptop3 的 MAC 地址（0060.5c74.2c99）。

②封装完成后，Laptop2 先将数据帧发送给交换机。

③交换机拿到数据帧以后，去读取数据帧上的目的 MAC 地址。

④交换机根据数据帧上的目的 MAC 地址查交换机缓存的 MAC 地址表，确定了 MAC 地址为 0060.5c74.2c99 的主机在端口 Fa0/13 上，于是就将数据帧从端口 Fa0/13 转发出去。

⑤Laptop3 就得到了 Laptop2 发来的数据帧，整个通信完成。

这个过程跟快递员送包裹的方式类似。快递员拿到快递后，也要看包裹上的目的地址，然后根据这个目的地址去送。

2. 学习数据帧的源 MAC 地址，填充 MAC 地址表

对于交换机而言，MAC 地址如此重要，所有数据帧的转发都要依赖于这个 MAC 地址表。那么这个表是怎么来的呢？是不是一开始就有的呢？

不是！交换机中的 MAC 地址表是设备启动后临时生成的，保存在缓存中，一开始只是一个空表。表中的记录是交换机在后续转发数据的时候一条一条学习得到的。

交换机如何学习 MAC 地址记录，不断地去填充 MAC 地址表？

当交换机从某个端口获得一个数据帧，会去读取数据帧上的源 MAC 地址，就知道这个端口是哪台主机在使用，其 MAC 地址是多少。然后将该 MAC 地址和对应交换机的端口号添加到 MAC 地址表中，形成 MAC 地址表中的一条记录。在本任务的案例中，一开始交换机并不知道 Fa0/12 端口是被哪台设备占用，更不知道该设备的 MAC 地址是多少。直到某一刻，交换机从 Fa0/12 端口接收到了一个数据帧，然后就去读取数据帧上的源 MAC 地址，就知道了连接到 Fa0/12 端口的设备的 MAC 地址是 0001.42d1.e839。

3. 数据帧的泛洪

如果交换机在 MAC 地址表中无法查找到目的 MAC 地址的记录，该如何处理呢？

既然 MAC 地址表一开始是空的，如果交换机拿到一个数据帧以后，根据数据帧上目的 MAC 地址去查 MAC 地址表，一定无法找到该目的 MAC 地址相应的记录，因此也就无法确定目标主机在哪个端口上。此时交换机会如何处置收到的这个数据帧呢？

答案是：泛洪！既然不知道目的主机在哪个端口上，那么采用类似"广播"的方式，将数据帧复制很多份，每个端口发一份，总有一个端口是对的吧。即使一个端口都不对，损失的资源也不多。

这种泛洪的方式是不得已而为之的，因为会带来一些安全问题。首先会占用交换机的 CPU 和缓存的资源，但这不是最主要的。风险更大的是，泛洪行为将数据帧发到每个端口，会给在网络中的嗅探者提供非法获取数据的机会，从而导致信息泄露。同时，黑客有可能会利用工具伪造不同 MAC 地址的无效的数据帧，不断发给交换机，交换机学习到新的记录以后，会将记录写入 MAC 地址表，直到 MAC 地址表被填满。真正的用户数据发来的时候，交换机学习到新的有效的 MAC 地址记录，但是已无法写入已经爆满的 MAC 地址表中。之后如果有数据帧要发给该 MAC 地址的主机，交换机就无法在 MAC 地址表中找到目的 MAC

地址的记录。于是只能通过泛洪的方式转发数据帧，我们可以用交换机端口安全的相关技术去解决这个问题。

课堂练习：用自己的语言描述交换机的工作过程。

素质拓展：交换机根据 MAC 地址表作出转发数据帧的决定：当在 MAC 地址表中找到目的 MAC 地址的记录，则按记录的端口转发出去；如果在 MAC 地址表中找不到目的 MAC 地址的记录，则采用泛洪的方式将数据帧发出。如果忽略安全的因素，那么泛洪的方式尽可能地减少了数据帧丢失的问题。我们在学习和将来的工作中，不可避免地会面临各种困难和挫折，要尽可能地去做更多尝试而不轻言放弃。而只有树立科学崇高的理想信念，才会给我们提供前进的动力，才会增加我们面对挫折的勇气和克服困难的信心。

任务 4：路由器参与数据通信——网络层

任务目标：理解路由器（三层交换机）的工作过程。

路由器（三层交换机）工作在 OSI 参考模型的网络层，负责处理 IP 数据包，也是计算机网络中使用最多的设备之一。因为其负责的是 OSI 参考模型的第三层，所以有时候习惯性地被称为三层设备。

路由器参与数据通信——网络层

1. 路由器转发 IP 数据包的过程

如图 4.5 所示，R0 是公司网络的边界路由器，左边连着公司内网，右边连着运营商的网络。

图 4.5 公司网络接入 ISP 网络拓扑

路由器负责的工作也很简单，就是将 IP 数据包从某个端口接收进来，然后从另一个端口转发出去。问题是：路由器也有很多个端口，它又怎么知道要从哪个端口将 IP 数据包转发出去，才能将 IP 数据包送到接收方？

答案是：查路由表。路由器的缓存中也有一个表，可以在特权模式下用 show ip route 命令查看，如图 4.6 所示。

当路由器接收到 IP 数据包，会去读取 IP 数据包上的目的 IP 地址，然后根据获取的目的 IP 地址去查路由表，根据子网掩码的最长匹配原则，找到相应的路由记录，最后决定通过路由记录中指定的端口转发出去。

比如路由器接收到一个 IP 数据包，发现目的地址是 192.168.0.1，则根据这个地址去查自己的路由表，匹配到表中第二条记录——192.168.0.0 的路由条目，然后将 IP 数据包从 FastEthernet0/0 端口转发出去。这个过程也和快递员投递包裹的过程类似，都是根据包裹上的目的地址来决定如何投递。

```
R0#sh ip route
Codes: C - connected, S - static, I - IGRP, R - RIP, M - mobile, B - BGP
       D - EIGRP, EX - EIGRP external, O - OSPF, IA - OSPF inter area
       N1 - OSPF NSSA external type 1, N2 - OSPF NSSA external type 2
       E1 - OSPF external type 1, E2 - OSPF external type 2, E - EGP
       i - IS-IS, L1 - IS-IS level-1, L2 - IS-IS level-2, ia - IS-IS inter area
       * - candidate default, U - per-user static route, o - ODR
       P - periodic downloaded static route

Gateway of last resort is 115.0.0.2 to network 0.0.0.0

     115.0.0.0/29 is subnetted, 1 subnets
C       115.0.0.0 is directly connected, FastEthernet1/0
     192.168.0.0/30 is subnetted, 1 subnets
C       192.168.0.0 is directly connected, FastEthernet0/0
S    192.168.10.0/24 [1/0] via 192.168.0.1
S    192.168.20.0/24 [1/0] via 192.168.0.1
S*   0.0.0.0/0 [1/0] via 115.0.0.2
```

图 4.6　路由器 R0 的路由表

值得注意的是，路由器将数据包转发出去以后，任务就结束，至于对端是否能收到，它不用管。甚至目的主机是否存在，它也不理会。只要在路由表中匹配到相应的路由条目，就根据路由条目标记的端口号去转发。

2. 路由器丢弃 IP 数据包

如果在路由表中匹配不到相应的路由条目怎么办？

对于路由器而言，要转发一个 IP 数据包，它会根据数据包上的目的 IP 地址去查路由表，如果匹配不到相应的路由条目，则将该 IP 数据包丢弃。这一点和交换机不一样，交换机在 MAC 地址表中如果找不到目的 MAC 地址对应的记录，会采用泛洪的方式，将数据帧发到每一个端口。

但是路由器在丢弃 IP 数据包之后，有一个细节显得特别"绅士"。路由器会给发送方发一个 ICMP（Internet Control Message Protocol，Internet 控制报文协议）包，告诉发送方：因为我这里匹配不到路由条目，你的 IP 数据包被我丢弃了！

3. IP 数据包进出路由器的细节

习惯上，我们说路由器工作在 OSI 参考模型的网络层，但其实它要完成的是包括网络层以下的所有三层的工作。数据包进出路由器的过程如图 4.7 所示。

路由器处理数据的整个过程如下：

①路由器从物理层接收二进制比特流，将得到的二进制比特流转换为数据帧。

②路由器的物理层将数据帧往上传输给数据链路层。

③路由器的数据链路层得到数据帧后，读取数据帧上的目的 MAC 地址，判断是否有权接收（如果 MAC 地址是自己的，或者是这个目的 MAC 地址是二层广播地址，则接收该帧），然后将数据帧解封，得到 IP 数据包，最后数据链路层将 IP 数据包往上传，给网络层。

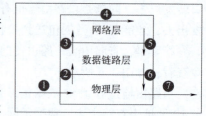

图 4.7　数据包进出路由器的过程

④路由器的网络层拿到 IP 数据包以后，读取数据包上的目的 IP 地址，根据这个 IP 地址去查路由表的记录，如果匹配不到，则将数据包丢弃。

⑤如果匹配到了，就根据匹配到的路由条目，决定要从哪个端口发出，但是端口工作在物理层，于是将IP数据包往下发给数据链路层。

⑥路由器的数据链路层得到IP数据包以后，将IP数据包封装成新的数据帧，并且将新的源MAC地址（自己的转出端口的MAC地址）和新的目的MAC地址（下一跳节点的MAC地址）写到数据帧上，最后将新的数据帧往下发给物理层。

⑦路由器的物理层得到数据帧后，将数据帧转换成二进制比特流，通过网络端口转发出去，给下一个节点。

路由器用于连接不同的网络，而广播包被限制在网络内传输，因此路由器很好地隔离了广播流量。

课堂练习：用自己的语言描述路由器的工作过程。

素质拓展：路由器根据路由表中的路由条目决定如何转发数据：匹配到路由条目就转发，没有匹配到就丢弃。丢弃数据包是为了防止一些数据包占用过多的网络资源，从而影响网络正常通信。有时候，放弃也不一定是坏事。我们不应该拘泥于眼前的得失，要树立正确的大局观和得失观，正确认识和对待人生发展过程中的得与失这对矛盾，走好人生之路，实现人生价值。

任务5：TCP确保可靠的数据传输——传输层

任务目标：理解TCP是如何保证数据的可靠传输的。

要深入探索OSI参考模型的传输层，就要先观察传输层在OSI参考模型中的位置，如图4.8所示。

TCP确保可靠的数据传输——传输层

传输层介于会话层和网络层之间，为会话层提供服务，依赖于网络层进行数据传输。当主机有数据要发送，则数据从应用层一路往下传，一直到传输层。传输层要负责的工作包括：对大的数据进行分片，将数据片封装成数据段，在数据段的头部填充信息，包括源端口号和目的端口号。

端口号是什么？这里说的端口号不是设备物理端口的编号，而是系统的服务或是某种应用程序在参与网络通信时被分配的编号，是一个抽象的概念。也就是说，端口号是标记一台主机上的不同应用或服务。数据段上的源端口号标记了数据是从发送方的哪个应用程序发出来的，而目的端口号标记的是数据要发到接收方的哪个应用程序。

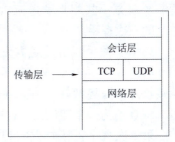

图4.8 传输层在OSI参考模型中的位置

传输层用端口号来标记应用层参与网络通信的应用程序或者服务。发送方的传输层负责标记数据从主机的哪个应用程序发出来，接收方的传输层从网络层拿到数据段后，根据目的端口号将数据发给相应的应用程序。至于自己属于哪台主机，数据准备发往哪台主机，这个任务交给网络层负责，传输层不管。

根据本项目任务2的分析结果，数据通信双方的传输层主要负责的工作内容是不一样的。发送方的传输层主要负责的工作包括：对数据进行分片，再将数据片封装成数据段，在数据段的头部填写各种参数,包括源端口号和目的端口号。接收方的传输层主要负责的工作包括:将获得的数据段进行解封装,得到数据片,对数据片进行重组,得到最初完整的数据,并根据目的端口号将数据发给相应的服务或应用。

为了完成 OSI 参考模型规定的传输层的工作,我们在传输层设计了很多协议,但主要是传输控制协议(Transmission Control Protocol,TCP)和用户数据报协议(User Datagram Protocol,UDP)。这两个协议都能独立完成传输层规定的任务,TCP 和 UDP 不需要彼此协作。

TCP 和 UDP 的主要区别在于提供的数据传输的质量不一样。TCP 是面向连接的,更能确保数据传输的可靠性;而 UDP 是面向非连接的,为上层提供的是不可靠的,尽力而为的数据传输服务,它不保证能将数据顺利地送给接收对方。简单地说,发送方如果将数据交给 TCP 负责传输,则发送方的 TCP 会事先和接收方的 TCP 进行同步(面向连接),达成一致以后才开始传数据。但是发送方如果将数据交给传输层的 UDP 负责传输,则情况会完全不一样。发送方的 UDP 会很努力地发送数据,但是它不会事先去和接收方的 UDP 同步(面向非连接),它不关心接收方是否做好了接收数据的准备,甚至接收方是否在网络中存活,它是不管的。只要上层有数据需要发送,它只管发送,不保证数据能到达对方,显然这样的传输方式是不可靠的。

这里来具体分析一下 TCP,它主要采取以下 4 种措施以确保通信双方数据的可靠传输。

1. 面向连接(三次握手)

面向连接就是在数据传输之前,发送方和接收方相互发送消息,实现同步,确定收发双方都已经准备就绪。如图 4.9 所示是 TCP 三次握手过程。

TCP 的收发双方要取得通信前的同步,需要相互发送自己的同步序列编号(Synchronize Sequence Numbers,SYN)消息给对方,同时在收到对方的消息后,要返回确认(Acknowledge,ACK)消息给对方。这个过程需要双方一共相互发送三次消息,被称为 TCP 的三次握手。

第一次握手:由发送方发起,发送一个同步消息 SYN 给接收方,表示想要发送数据给接收方,提醒做好准备。同步消息的"序号"SEQ 值为一个随机数,这里是 100。

第二次握手:接收方收到发送方发来的同步消息,按规则必须要给发送方一个确认

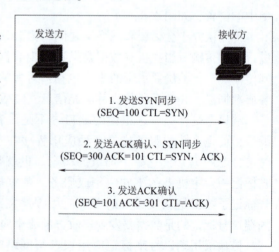

图 4.9　TCP 三次握手过程

ACK(ACK 的值为收到的 SYN 的 SEQ 值加 1,这里第一次握手的时候发送的 SEQ 值为 100,而接收方为了表示已经接收到这个 SYN 消息,则将 SEQ 值加上 1 的值 101 发回给发送方),表示消息已经收到。同时也要将自己的 SYN 消息发送给发送方,此时发出的 SEQ 值也是随机的,这里是 200。

第三次握手:发送方收到接收方的第二次握手的同步 SYN 消息后,按规则要给予确认,

ACK 的值也等于第二次握手的 SEQ 值加 1（这里是 301）；此时的 SEQ 值不再是随机的，而是第二次握手的 ACK 值。

三次握手完成后，发送方和接收方完成同步（准备就绪），即接收方准备好接收，发送方准备好发送，于是通信可以开始。

2. 数据的分片、重组机制

发送方的传输层在从上层接到数据以后，首先做的一个事情就是将用户数据进行分片，并且将数据片进行封装，得到数据段。然后给这些数据段进行编号，将编号填写在数据段的头部。

发送方将数据进行编号的作用是，方便接收方在收到一组数据段后进行重组的工作。保证数据按原来的顺序再组建回来，保证数据不会出错。

在错综复杂的网络中，数据的传输并不都能按理想的方式到达。同一组数据不同编号的数据段在传输的过程中由于可能经过的路径不一样，导致先发送的数据段可能比后面发送的数据段还要迟。发送方按数据段的编号顺序发送，但是到达发送方的顺序有可能会变化。为了保证数据的完整性，接收方的 TCP 协议会先按数据段的编号进行排序，再进行重组，无差错得到原始数据，如图 4.10 所示。

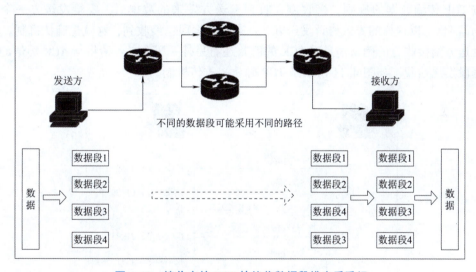

图 4.10　接收方的 TCP 协议将数据段排序后重组

3. 滑动窗口的流量控制机制

TCP 同时提供流量控制机制，即接收方能够可靠地接收并处理的数据量。流量控制可以调整给定会话中源和目的地之间的数据流速，有助于保持 TCP 传输的可靠性。为此，TCP 报头包括一个称为"窗口大小"的 16 位字段。

初始窗口大小在三次握手期间建立 TCP 会话时确定。发送方必须根据目的设备的窗口大小限制发送到接收方的字节数。只有发送方收到数据段已接收的确认（ACK）消息之后，才能继续发送更多数据段。通常情况下，接收方不会等待其窗口大小的所有字节接收后，才确认应答。接收和处理字节时，接收方就会发送确认，以告知发送方它可以继续发送更多字节。

接收方在处理接收的数据段，并不断调整向发送方返回确认窗口大小，这个过程被称为滑动窗口。通常，接收方在每收到两个数据段之后发送确认。在确认之前收到的数据段的数量可能有所不同。滑动窗口的优势在于，只要接收方确认之前的数据段，就可以让发送方持续传输数据段。

接收方一般会根据自己缓存的大小和数据段的丢失率来调整窗口大小。如果自己的缓存已经很小了，则发送方会将窗口调小，告诉发送方减小发送的数据段的量，以免造成接收方因缓存不够而将收到的数据段丢弃，反之，如果接收方还有很大的缓存，则会将窗口调大，然后通知发送方；如果接收方接收到的数据段丢失率很高，则其判断可能是由于传输的网络出现阻塞，网络设备为了避免拥塞，丢弃了一部分数据，此时接收方也会将窗口减小，告诉发送方减少发送的数据段的数量，以减小传输通道上流量，这在一定程度上缓解了网络路径的阻塞，提高数据传输的成功率；反之，如果接收方发现数据的丢失率很低，则会将窗口调大，然后将窗口大小通知发送方。

4. 确认、重传机制

按 TCP 传输数据的规则，接收方在收到发送方发来的数据后，要给发送方一个确认（ACK）消息，以这样的方式告诉发送方，发来的数据段已经收到，这就是确认机制。

发送方给接收方的确认消息（ACK 的值）还有另外一层含义：请将与 ACK 等值的序号的数据段发送给我。如图 4.11 所示是 TCP 确认、重传机制。

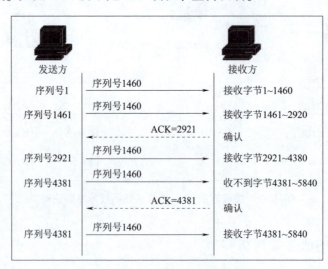

图 4.11　TCP 确认、重传机制

由于网络的复杂性和环境的影响，数据在传输的过程中难免会出现差错或者丢失的情况，确认和重传机制在很大程度上确保了数据传输过程中的完整性。

课堂练习：用自己的语言描述 OSI 参考模型传输层中 TCP 和 UDP 工作方式的异同。

素质拓展：TCP 为了确保数据的可靠传输，主要采取了三次握手、数据的分片和重组、滑动窗口的流量控制、确认重传等机制。TCP 是一个可靠的、负责任的协议。我们也要培养自己的责任意识。学会对自己负责，对亲人负责，对周围的人和更多的人负责，进而对民族、国家、社会负责，做一个有价值、负责任的人。要正确认识和处理人生中遇到的各种问题，不能得过且过、放纵生活、游戏人生，否则就会虚掷光阴，甚至误入歧途。

小结与拓展

1. 比较 OSI 参考模型和 TCP/IP 模型

网络模型是计算机网络的一种抽象的表达，ISO 组织发布的计算机网络模型叫 OSI 参考模型，分 7 层。但是还有一种更简洁，同样也被业界认可和接受的模型，叫 TCP/IP 模型。这两个模型的对应关系，如图 4.12 所示。

计算机网络的作用就是实现数据传输，而网络模型只是计算机网络的一种抽象的描述。描述的方式可以有很多种，同时也可以提出很多种网络模型，但是不管怎样去描述，计算机网络的功能是不变的。也就是说，不管用什么模型来描述计算机网络，这些模型所定义的网络的功能是一样的。

TCP/IP 模型和 OSI 参考模型只是对计算机网络的不同描述，而这些描述并不会影响到计算机网络的本质——实现数据传输。

区别在于，TCP/IP 模型对计算机网络的描述更简洁。TCP/IP 模型将 OSI 参考模型中上三层（会话层、表示层、应用层）的功能统一为一层来描述，即应用层，下两层（物理层、数据链路层）的功能统一为一

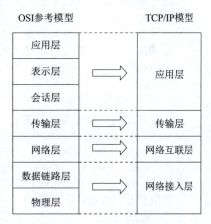

图 4.12　OSI 参考模型和 TCP/IP 模型的对应关系

层来表述，即网络接入层，中间的传输层依然对应于传输层，网络层也对应与另外一层，但换了个名字，叫网络互联层。

OSI 参考模型用了 7 层描述计算机网络，而 TCP/IP 模型用了 4 层描述了计算机网络，但并不意味着 TCP/IP 模型所要完成的工作量就少。其实它们所要完成的工作量是一样的，只不过对工作进行了重组。这就像是一个公司对各业务部门进行了重组，但不管怎么组合，公司要正常运转，总的工作量还是不会减少或增多。

2. 比较 TCP 和 UDP

传输层有两个主要协议，一个是 TCP，另一个是 UDP，这两个协议是并行的，是不相干的，它们各自都能独立完成传输层规定的工作。那么它们有什么区别呢？既然都能独立完成，为什么要在传输层设计两个协议而不是只开发一个协议就可以了呢？

区别是：TCP 是一个可靠的协议，UDP 是一个不可靠的协议。传输重要数据的时候，我们会更多地考虑将数据交给 TCP 负责传输，而对于一些不是很重要，对延迟要求比较高的数据，就可以考虑交给 UDP 去传输。

TCP 之所以可靠，在前面我们已经分析过，那相对于 TCP 而言，UDP 究竟不可靠到什么程度呢？我们通过表 4.3 将这两个协议做一个对比。

表 4.3 TCP 和 UDP 对比

序号	TCP	UDP	备注
1	面向连接：TCP 在传数据之前，发送方和接收方的 TCP 要通过"三次握手"取得同步，准备好数据的发送和接收	面向无连接：UDP 发送数据的时候，不会进行同步。发送方的 UDP 只管发送，它不会关心接收方是否做好了接收的准备，甚至不管接收方是否存在	TCP 通过"三次握手"在收发双方建立了虚电路，占用了时间以及其他开销；数据传输开始之前 UDP 不需要任何开销
2	确认重传：接收方的 TCP 接收到数据以后，会给发送方一个确认消息，表示数据段已经收到；如果数据在传输的过程中被损坏或丢失，则接收方也会通过 ACK 消息要求发送方再次发送对应序号的数据段	无确认：接收方的 UDP 在接收到数据段以后，不会给发送方的 UDP 发送确认消息。收发双方就像两个生闷气的小孩，你发送的时候没通知我，那我接收到的时候也不想告诉你	TCP 通过确认重传的方式确保数据无差错达到对端，增加了工作量，影响数据传输效率；UDP 没有任何措施，数据传输可能会出错，但是效率很高
3	排序重组：发送方的 TCP 在将数据进行切片的时候，对不同的数据段进行编号；当数据段到达目的地的时候，接收方的 TCP 会根据数据段的编号进行重组，确保数据的完整性	无序重组：发送方的 UDP 在将数据进行切片的时候，是没有编号的。封装好数据段以后，就发送出去；显然，大多数时候，数据到达接收方的顺序是乱的，此时接收方的 UDP 按数据到达的顺序进行重组，因为发送方没有对数据段进行编号，所以无法按编号重组，会导致数据混乱	接收方的 TCP 必须预留更多缓存，存放同一组数据的所有数据段，必须等到所有数据段到达后，才可以排序重组；如果用 UDP 来传输，则接收方按先到先重组的方式进行，不需要太多缓存，同时工作效率也得到了提高
4	滑动窗口流量控制：TCP 的双方在第三次握手的时候，确定好了窗口的大小，接收方会根据自身缓存大小和数据段的丢失率来调整窗口大小，并通知发送方。发送方会根据窗口大小来发送数据，以缓解可能的网络阻塞，确保数据传输的成功率	没有流量控制机制：UDP 在有数据需要发送的时候，努力（以最大可能的数据量）去发送，不管网络是否出现阻塞，不进行任何的控制，有可能会导致网络更拥堵而产生雪崩效应	
5	报头大小：20 Byte	报头大小：8 Byte	TCP 报头比 UDP 报头大，则开销大

从表中的对比可以看出，UDP 虽然不那么可靠，但开销小，传输数据的效率比 TCP 要高。TCP 的确可以保障数据传输的可靠性，但是也付出了很多的资源开销，影响数据传输效率。实际上，以前计算机网络设备制造工艺不是很精的时候，网络设备处理数据的确会带来一些数据丢失的现象，但是随着工业工艺技术水平的提高，现在的网络数据传输质量相对于计算机网络诞生早期已经有了很大的提高，所以很多时候，我们可以不用去考虑太多的因素，比如要借助 TCP 来保障数据的可靠传输等。

3. 交换机转发数据的两种方式：广播和泛洪的区别

交换机对数据帧进行广播或泛洪，其行为都是一样的：将数据帧复制 N 份，每个活动端口发一份出去。区别在于，广播帧的目的 MAC 地址是一个广播地址，泛洪帧的目的 MAC 地址是目的主机的地址，是特定的某台设备的 MAC 地址。

以太网是一种广播型的网络，数据链路层在封装以太网帧的时候，要将源 MAC 地址和目的 MAC 地址写入帧的头部。源 MAC 地址就是发送数据主机自己的 MAC 地址，这个是可以确定的。

但是当发送方要发送一个帧给网络中的所有主机时，则会在数据帧头部的目的 MAC 地址字段写入二层广播地址（全 F），交换机转发这类帧的行为叫广播。

当交换机接到一个数据帧，然后根据这个数据帧上的目的 MAC 地址去查询 MAC 地址表却在表中找不到时，则采用泛洪的方式将数据帧发出。

4. ARP 查询

先分析发送方和接收方在同一个网络中的情况。比如：发送方要发送一组数据给主机 192.168.0.1。

在发送方，当数据一路往下，到数据链路层准备要进行第三次封装时，需要在数据帧头部填写源 MAC 地址和目的 MAC 地址。源 MAC 地址是自己的地址，发送方自己是知道的。但是要填写目的 MAC 地址的时候，发送方却不知道目的主机的 MAC 地址。

为了完成帧的封装，发送方必须询问接收方的 MAC 地址，可是又不知道目的主机是哪台设备，因此发出一个广播帧（帧的目的地址是全 F），这个广播帧就是 ARP 查询。其内容是：请问谁是 192.168.0.1，请把你的 MAC 地址发给我。目的主机收到请求以后，返回一个应答，将自己的 MAC 地址告诉发送方。发送方得到目的主机的 MAC 地址后，数据链路层数据帧的封装得以继续。

因为 ARP 查询是以广播的方式发出来的，所以如果发送方和接收方不在同一个网络，收发双方直接的 ARP 查询则无法实现，因为 ARP 广播帧无法跨网络传输。数据帧只负责网络内的数据传输，跨网络的环境下，数据是一跳一跳往下传输的。每经过一个网络，数据链路层必须封装新的帧，以便将数据帧发送给下一跳设备。如图 4.13 所示是不同网络间的 ARP 查询。

图 4.13　不同网络间的 ARP 查询

发送方和接收方不在同一个网络，则 ARP 解析的过程如下：
①发送方先将数据发给网关（R1）。
②R1 将自己的 MAC 地址返回给发送方，发送方完成数据帧的封装以后，将数据发送给 R1。

③R1 拿到数据帧以后，对数据帧进行解封装，得到 IP 数据包以后发送到网络层进行路由查询。R1 完成路由查询后，决定要将数据转发给 R2，于是 R1 通过 ARP 查询的方式，询问 R2 的 MAC 地址。

④R2 将自己的 MAC 地址返回给 R1。R1 的数据链路层根据查询得到的 R2 的 MAC 地址封装新的数据帧，然后将新的数据帧发送给 R2。

⑤R2 向接收方发出 ARP 查询。

⑥接收方将自己的 MAC 地址返回给 R2，由 R2 封装新的帧后发给接收方，完成最后一跳的数据传输。

5. 路由器的递归查询

递归，在数学与计算机科学中，是指在函数的定义中使用函数自身的方法。在路由表查询的案例中，指的是对路由表的多次查询，即根据前一次查询路由表得到的结果去进行第二次查询的方式。

路由器 R0 的路由表如图 4.14 所示。

```
R0#sh ip route
Codes: C - connected, S - static, I - IGRP, R - RIP, M - mobile, B - BGP
       D - EIGRP, EX - EIGRP external, O - OSPF, IA - OSPF inter area
       N1 - OSPF NSSA external type 1, N2 - OSPF NSSA external type 2
       E1 - OSPF external type 1, E2 - OSPF external type 2, E - EGP
       i - IS-IS, L1 - IS-IS level-1, L2 - IS-IS level-2, ia - IS-IS inter area
       * - candidate default, U - per-user static route, o - ODR
       P - periodic downloaded static route

Gateway of last resort is 115.0.0.2 to network 0.0.0.0

     115.0.0.0/29 is subnetted, 1 subnets
C       115.0.0.0 is directly connected, FastEthernet1/0
     192.168.0.0/30 is subnetted, 1 subnets
C       192.168.0.0 is directly connected, FastEthernet0/0
S    192.168.10.0/24 [1/0] via 192.168.0.1
S    192.168.20.0/24 [1/0] via 192.168.0.1
S*   0.0.0.0/0 [1/0] via 115.0.0.2
```

图 4.14　路由器 R0 的路由表

如果 R0 收到一个 IP 数据包，目的网络为 192.168.10.0，其查询路由表的过程是怎样的呢？

首先 R0 会根据 IP 数据包上的目的地址去匹配到路由表中的条目。

```
S   192.168.10.0/24[1/0]via  192.168.0.1
```

由返回信息可知，R0 决定要将数据包转发给下一跳 192.168.0.1，但是不知道这台设备是连接到 R0 的哪个端口。于是要再对路由器进行第二次查询，匹配到路由表中的另外一个条目。

```
C   192.168.0.0/24 is directly connected,FastEtherner0/0
```

从该路由条目得知主机 192.168.0.1 所在的网络连接到端口 FastEtherner0/0，最后才决定要将 IP 数据包从端口 FastEtherner0/0 发出去。

显然，为了将 IP 数据包转发出去，这里要查询两次路由表，后面的查询是以前面的查询结果为依据。

项目4 解析网络通信

6. IP 数据包的生命周期（Time To Live，TTL）

TTL 值是 IP 数据包报头的一个字段（8 bit），用于限制数据包的生命周期。

IP 数据包的初始 TTL 由数据包的源设备负责设置，当数据包被路由器（三层设备）处理一次，TTL 值就减 1。如果 TTL 字段的值减到 0，则三层设备将丢弃该数据包，并向源设备发送一个 ICMP 超时消息。告诉 IP 包的发送方，数据包的生命周期已经结束，但还找不到目的主机，该数据包已经被丢弃。

7. 可变长子网掩码（VLSM）

先回顾一下 IP 地址。我们已经知道，IP 地址用于识别互联网中不同的主机。用子网掩码来确定 IP 地址中哪些位是主机位，哪些位是网络位。主机位的位数，决定了该网络中能表达的不同的 IP 地址数量。假设某个 IP 地址的主机位是 n 位，根据数学组合，容易算出该网络中最多有 2^n 个不同的 IP 地址。在这些 IP 地址中，最小的地址叫网络地址，用于标识不同的网络，最大的地址叫广播地址，用于封装广播数据包。剩下的 2^n-2 个地址可以分配给网络内的主机，用于通信。

当一个网络的主机数量很多，IP 地址不够用时，可以允许主机位向网络位借位，增加主机位的位数，实现 IP 地址扩容。同样的，如果需要的 IP 地址不用这么多，不想导致 IP 地址浪费，则可以从主机位借出子网位，减小网络的 IP 地址容量。

再了解固定长度的子网掩码。如果一个公司申请了一个 C 类的地址：211.1.1.0/24，公司有 4 个部门，需要划分 4 个子网。于是主机位可以借 2 个位出来作为子网位，此时网络位的总长度不再是 24，而是 26。如图 4.15 所示是将原有 IP 地址空间进行划分。

按照子网划分的结果，每个部门的 IP 地址空间为 64 个，减掉网络地址和广播地址，还有 62 个 IP 地址可以分配给主机使用，也就是说，每个部门最多可以容纳 62 台计算机（包括网关）。

部门A:	11010011.00000001.0000001.	00	000000	/26
部门B:	11010011.00000001.0000001.	01	000000	/26
部门C:	11010011.00000001.0000001.	10	000000	/26
部门D:	11010011.00000001.0000001.	11	000000	/26
	网络位	子网位	主机位

图 4.15 将原有 IP 地址空间进行划分

但是如果考虑各部门的人数，那么采用固定长度子网掩码的方式划分子网，会导致 IP 地址的浪费。如表 4.4 所示是各部门的 IP 地址数量统计表。

表 4.4 各部门的 IP 地址数量统计表

部门	计算机（台）	分配的主机地址数量	浪费的地址数量
A	2	62	60
B	28	62	34
C	10	62	52
D	50	62	12

显然，对于 A、B、C 三个部门而言，浪费了很多 IP 地址。可不可以再优化 IP 地址的划分，以节约更多 IP 地址，用于今后网络的扩展呢？

采用可变长子网掩码，可以最大限度避免 IP 地址的浪费。可变长子网掩码划分子网的思路和计算步骤如表 4.5 所示。

表 4.5　可变长子网掩码划分子网的思路和计算步骤

步骤	项目	部门 A	部门 B	部门 C	部门 D	备注
1	计算机数量	2	28	10	50	确定计算机数量
2	需要的主机地址数	2	28	10	50	每台计算机需要 1 个 IP 地址
3	需要的 IP 地址数	4	30	12	52	主机地址加 2 个特殊地址（网络地址和广播地址）
4	分配的 IP 地址数	$2^2=4$	$2^5=32$	$2^4=16$	$2^6=64$	不能任意分配。只能按 2^n 个来分配，以刚好满足需要为准。n 为主机位数
5	每个子网的 IP 地址范围	211.1.1.112 ~ 211.1.1.115	211.1.1.64 ~ 211.1.1.95	211.1.1.96 ~ 211.1.1.111	211.1.1.0 ~ 211.1.1.63	按 IP 地址需求量从大到小分配，这里按 D—B—C—A 的顺序分配
6	子网掩码长度	/30	/27	/28	/26	子网掩码长度为：32-n
7	网络地址	211.1.1.112/30	211.1.1.64/27	211.1.1.96/28	211.1.1.0/26	子网中最小的 IP 地址
8	广播地址	211.1.1.115/30	211.1.1.95/27	211.1.1.111/28	211.1.1.63/26	子网中最大的 IP 地址
9	可用的主机地址范围	211.1.1.112/30 ~ 211.1.1.115/30	211.1.1.64/27 ~ 211.1.1.95/27	211.1.1.9/28 ~ 211.1.1.111/28	211.1.1.0/26 ~ 211.1.1.63/26	子网中 IP 地址段的中间范围

注意：分配 IP 地址的时候，要先分给大网络，再分给小网络，否则会出现 IP 地址碎片化，不利于将来网络的扩展。

8. 端口号基础知识

传输层的端口号用于区分同一台主机上不同的应用或服务。互联网数字分配机构（Internet Assigned Numbers Authority，IANA）是负责分配各种编址标准（包括端口号）的标准机构。端口号有以下不同类型。

公认端口（端口 0 到 1023），这些编号用于服务和应用程序。Web 浏览器、电子邮件客户端以及远程访问客户端等应用程序通常使用这些端口号。通过为服务器应用程序定义公认端口，可以将客户端应用程序设定为请求特定端口及其相关服务的连接。

注册端口（端口 1024 到 49151），这些端口号由 IANA 分配给请求实体以用于特定进程或应用程序。这些进程主要是用户选择安装的一些应用程序，而不是已经分配了公认端口号的常用应用程序。例如，思科已将端口 1985 注册为其热备份路由器协议（Hot Standby Router Protocol，HSRP）进程。

动态或私有端口（端口 49152 到 65535），也称为临时端口。这些端口通常是在主机向服务器发起连接时由客户端操作系统动态分配。动态端口用于在通信期间识别客户端应用程序。

常见应用或服务的端口号如表 4.6 所示。

表 4.6 常见应用或服务的端口号

端口号	传输层协议	应用/服务	缩写词
20	TCP	文件传输协议（数据）	FTP
21	TCP	文件传输协议（控制）	FTP
22	TCP	安全外壳	SSH
23	TCP	Telnet	——
25	TCP	简单邮件传输协议	SMTP
53	TCP、UDP	域名服务	DNS
67	UDP	动态主机配置协议（服务器）	DHCP
68	UDP	动态主机配置协议（客户端）	DHCP
69	UDP	简单文件传输协议	TFTP
80	TCP	超文本传输协议	HTTP
110	TCP	邮局协议第 3 版	POP3
143	TCP	Internet 消息访问协议	IMAP
161	UDP	简单网络管理协议	SNMP
443	TCP	安全超文本传输协议	HTTPS

了解一些常见应用或服务的端口号非常重要，首先是网络管理工作经常会涉及，其次是设计网络访问策略的时候，避免影响正常应用或服务。

9. 套接字

为了清楚表达一个数据通信的双方，我们要同时借助 IP 地址和端口号这两个概念。

这里要把 IP 地址和端口号做一个区分。IP 地址标记的是网络中不同的主机，端口号则标记一台主机中参与网络通信的不同的应用程序或服务。源 IP 地址标记了数据从哪一台主机发出来，而目的 IP 地址标记的是数据要发到网络中的哪一台主机。源端口号标记了数据从源设备上的某个应用程序发出来，目的端口号则表示数据要发到目的主机的哪个应用程序。

套接字（Socket）将 IP 地址和端口号整合在一起，描述了完整的通信双方。套接字用 IP 地址和端口号来一起表述，格式为："IP 地址：端口号"，描述了某台主机上的某个参与通信的应用程序或服务。用套接字来详细表示一个数据通信：主机 192.168.0.1 上端口号为 8080 的应用程序，发送一组数据给主机 192.168.0.2 上端口号为 80 的应用程序或服务。如图 4.16 所示是用 Socket 描述数据通信过程。

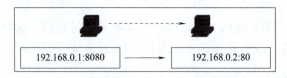

图 4.16　用 Socket 描述数据通信过程

用套接字描述数据通信比单独用 IP 地址描述主机之间的通信或单独用端口号描述不同应用程序之间的通信更详细、更准确。因为仅用 IP 地址描述通信双方，只能描述数据从哪台主机发往哪台主机，而传输的这些数据来自主机上的哪个应用，无法表达；如果单独用端口号来表述通信双方，则只能描述数据是哪个应用程序发出来的，要发给哪个应用程序，但无法描述应用程序是哪台主机上的。

10. TCP 的三次握手——协同攻击难题

发送方和接收方的 TCP 协议在传输数据之前实现同步，相互之间通信的次数为什么是三次？而不是两次、四次或其他？

这来自一个经典的博弈论案例：协同攻击难题——将军的困境。

故事说的是两位将军，各带领自己的部队埋伏在相距一定距离的两座山上，等候敌人。将军 A 得到可靠情报说，敌人刚刚到达，立足未稳。如果两股部队一起进攻敌军，就能够获得胜利；而如果只有一方进攻，进攻方将失败。

将军 A 遇到了一个难题：如何与将军 B 协同一致、一起进攻？

那时没有电话之类的通信工具，而只能通过情报员来传递消息。将军 A 派遣一个情报员去告诉将军 B：敌人没有防备，两军于"黎明一起进攻"。然而可能发生的情况是，情报员失踪或者被敌人抓获，即将军 A 虽然派遣情报员向将军 B 传达"黎明一起进攻"的信息，但他不能确定将军 B 是否收到他的信息。事实上，即使情报员回来了，将军 A 又陷入了迷茫：将军 B 怎么知道我的情报员肯定回来了？

将军 B 如果不能肯定情报员已经安全回来，他必定不会贸然进攻的。于是将军 A 又将该情报员派遣到将军 B 处。然而，他不能保证这次情报员肯定到了将军 B 那里。

对于这个问题，两位将军不管互派多少次情报员，都无法取得协同。这里我们将问题优化一下：假设环境是安全的，两位将军约定，如果收到对方的消息，只需要给对方确认即可。那么在优化后的理想条件下（环境安全，情报员没有遇到任何危险），两位将军互派情报员的情境如图 4.17 所示。

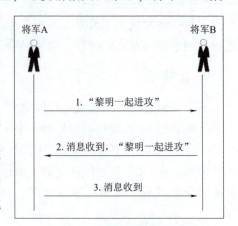

图 4.17　理想状态下两位将军互派情报员的情境

可见，理想状态下，两位将军至少需要三次通信，以取得协同。TCP 三次握手的思路是一样的：

发送方和接收方至少需要三次通信，才能实现同步。
对应于 TCP 的三次握手：
①发送方：我已经准备好发送（SYN）；
②接收方：消息收到（ACK），我已经准备好接收（SYN）；
③发送方：消息收到（ACK）。

11. TCP 三次握手带来的安全问题

如果接收方接到了发送方发来的 SYN 后，回了 SYN+ACK 给发送方，此时发送方离线了，接收方没有收到发送方返回来的 ACK，第三次握手失败，如图 4.18 所示。

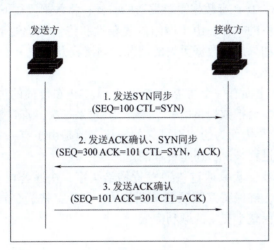

图 4.18　第三次握手失败

那么，TCP 的三次握手就处于一个中间状态，既没成功，也没失败。于是，接收方如果在一定时间内没有收到发送方发出的第三次握手消息，接收方的 TCP 协议会重发第二次握手消息（ACK+SYN）给发送方。对 Linux 系统而言，默认重试次数为 5 次，重试的间隔时间从 1 s 开始每次都翻倍。5 次的重试时间间隔为 1 s，2 s，4 s，8 s，16 s，一共 31 s；第 5 次发出后还要等 32 s 才知道第 5 次也超时了，所以，总共需要 1+2+4+8+16+32＝63 s，接收方的 TCP 才会断开这个连接，释放相关资源。

如果重试延时机制被恶意利用，发送方向接收方发出大量的 TCP 请求，但又故意不发出第三次握手的消息（ACK），会消耗接收方大量的资源，最终有可能导致接收方因资源耗尽而无法提供正常的服务。

Linux 系统可以使用 3 个 TCP 参数来调整重试延时的行为。_synack_retries 参数，调整重试次数；_max_syn_backlog 参数，调整 SYN 连接数；_abort_on_overflow 参数，调整超出连接能力时的行为。

12. TCP 是可靠的，但是它又将数据发给不可靠的 IP 负责传输，结果是可靠还是不可靠的呢

首先我们看一下 OSI 参考模型中 TCP 和 IP 的关系。TCP 工作在传输层，而 IP 工作在

网络层。如图 4.19 所示是 OSI 参考模型中 TCP 和 IP 的关系。

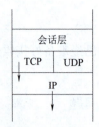

图 4.19 OSI 参考模型中 TCP 和 IP 的关系

IP 为上层提供服务，TCP 有数据需要发送的时候，将数据段往下发送给 IP，由 IP 发到目的主机。正如前面分析的一样，由于网络的复杂性和传输环境的干扰，数据在传输的过程中难免会出现差错。如果出现数据丢失的情况，则接收方的 TCP 会发送 ACK 消息，通知发送方的 TCP 再发送一次。

这就像不可靠的 IP 上面有一个可靠的 TCP 上司。IP 在传输的过程中把数据弄丢了不要紧，TCP 发现了以后，会把数据再给 IP，要求 IP 再发一次，直到数据成功到达接收方。

所以即使 IP 是一个尽力而为的，不可靠的协议，但是由于有一个可靠的 TCP 来监控，所以就数据传输的整个过程来说，还是可靠的。

类似于我们在线购物。卖家通过快递发货物给买家，快递公司有可能会把快递弄丢，此时，卖家会再发一份，确保买家能顺利拿到货物。所以，快递公司不可靠不要紧，只要卖家是可靠的，就能确保整个交易的顺利完成。

思考与训练

基于 OSI 参考模型，描述如图 4.20 所示的任务拓扑图中内网用户 Laptop1 访问 Internet 网络 Web 服务器的数据通信过程细节。

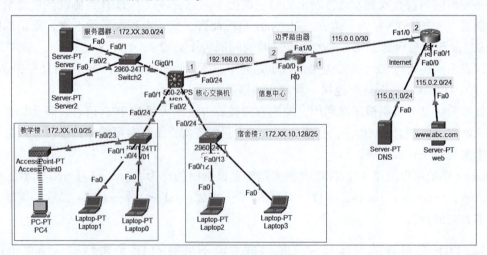

图 4.20 任务拓扑图

Project 4

Analyzing Network Communications

In terms of learning itself, it is important to not only "know what you know" but also "know why you know what you know" in order to be able to apply it flexibly and make sense of it. This project focuses on explaining how computers communicate with each other, and sublimates the knowledge level from practice to theory. By introducing the OSI reference model, it reinterprets what a computer network is from a microscopic point of view. The concept of layering is proposed, and how each layer of the OSI reference model collaborates with each other to complete data communication, clarifying every detail of the communication involving computers, switches, and routers, and laying a good foundation for subsequent troubleshooting.

Learning Objectives

(1) Understand commonly-used network models (OSI reference model and TCP/IP network model);
(2) Understand the role of the layers in the OSI reference model and the collaboration between layers in the data communication process;
(3) Understand the role of switches and routers in the communication process;
(4) Understand the meaning of VLSM (Variable Length Subnet Mask).

Skill Objectives

(1) Be able to describe the function of each layer of the OSI reference model;
(2) Be able to clearly describe a complete data communication process based on the OSI reference model;
(3) Master the VLSM approach to partition large networks into smaller ones.

Relevant Knowledge

Through Project 1, we have learnt what a computer network is. But in fact, that is from the macro, specific point of view to describe the computer network, which is easier for beginners to understand. But if we want to figure out how the network works, how the data is forwarded from the sender to the receiver step by step, and how each detail of the whole communication works, we have to describe what a computer network is from another perspective. Unlike the macro, concrete perspective, this perspective should be more abstract and specialised, but it does help us understand the data communication process better.

Some may wonder: why should we care about the details of data communication? As long as the network equipment can help the user to send the data to the receiver, the data transmission task should be completely taken by the network intermediary equipment. Here we need to sort out our future identity-network engineers, not ordinary network users. When the network has a variety of failures, network engineers should be in the shortest possible time to troubleshoot and restore network communication, which is like a doctor's duty is to treat the disease to save lives. But if this doctor is not clear about the details of the work of the organs of the human body and how the organs work together to protect the great project of life, I believe that almost no one dares to to see such a doctor. Similarly, if we do not understand the principle of network, do not understand the details of data communication, when the network fails, we can not judge where the point of failure is, what causes the failure, how to solve it, and how to avoid it in the future. Will any customer look for this kind of person who doesn't understand the principles of communication to help solve network communication problems? Hardly.

In order to explain more professionally what a computer network is, we introduce the concept of a network model.

A network model is simply a representation of how a computer network works and not the actual network. It describes what a particular layer needs to accomplish from the perspective of network functionality, but does not prescribe how it should be accomplished as a way of maintaining openness and consistency among the various types of network protocols and services.

There are many network models, two of the more commonly-used ones: the OSI reference model and the TCP/IP network model.

Task 1: A First Look at the OSI Reference Model

A First Look at the OSI Reference Model

Mission Objective: Understand the idea of layering the OSI reference model and the function of each layer.

When computer networks were first introduced, usually only computers made by the same manufacturer could communicate with each other. When an enterprise deployed a network, it had to purchase products from the same company and could not integrate equipments from several manufacturers. It was not until the late 1970s that the International Organisation for Standardization (ISO) developed the Open System Interconnect (OSI) reference model. The purpose of the model is to help manufacturers produce interoperable network equipments and softwares in the form of protocols that allow networks from different manufacturers to communicate with each other.

The OSI reference model is the primary network architecture model that describes how data and network information are transferred over the network medium from one computer application to another. In order to clearly describe the details of data transmission, the OSI reference model counts and groups (layers) all the work tasks of a single network communication process, each layer specifies what work needs to be done, and layers collaborate with each other to complete the entire communication task. Just like the logistics company in order to complete the express delivery, set up several departments, the provisions of each department should complete the task, each department has its own responsibilities and collaborates with each other, and ultimately completes the transmission of an express. Here the OSI reference model defines the layer, similar to the company setting up the business sector.

The OSI reference model defines seven layers: physical, data link, network, transport, session, presentation, and application, as shown in Figure 4.1.

The seven layers of the OSI reference model are divided into two groups Group 1: physical, data link, and network layers; and Group 2: transport, session, presentation, and application layers. Group 1 is responsible for end-to-end data transfer and Group 2 is responsible for communication between applications. In other words, Group 1 is responsible for data sent from this host to another host, and it does not care about which application in the host sends the data and which application in the opposite host receives it. This is similar to the courier company which is responsible for the transfer of

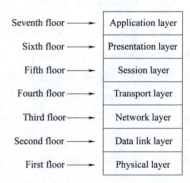

Figure 4.1 OSI Reference Model

goods to the truck driver, he will not care about the carriage of the courier is from where and send to whom, he is responsible for relaying relay the goods from the city to the city. Group 2 is responsible for running on the end host applications communicate with each other, but also to provide interaction with the end-users, this work is similar to the courier company is responsible for the task to complete the employees to receive the courier. This job is similar to that of an employee at a courier company who is responsible for receiving deliveries. Unlike the driver of a large truck, the courier has to determine the identity of the user who sends the courier, and the employee in charge of delivery is responsible for locating the person receiving the courier in the area and collecting the courier with a valid ID.

Reference model each layer of the work to be done is not the same, between the layers and to collaborate with each other, together to complete a data transmission. Just like the courier company's employees in different positions are responsible for the work are not the same, the positions of the staff to work together, any one role is indispensable, together to complete the transmission of express.

What does each layer of the reference model need to accomplish? The functions of each layer of the OSI reference model are listed below:

1. Application Layer

The application layer is the channel through which the user interacts with the computer and comes into play when access to the network is required. Protocols that work at the application layer include HTTP, DNS, DHCP, FTP, SMTP, and POP3.

However, there are some applications that do not work at the application layer, such as Internet Explorer. Even if you disable all network protocols and network cards, IE can still browse local files, but once it receives a user request for access to network resources, it is only when it goes to call the ports and services of the application layer. At this point, the service is the application layer procedures.

The application layer also has the important task of being responsible for the availability of the target communicator and determining whether there are enough resources to ensure the desired communication.

2. Presentation Layer

OSI defines at the presentation layer how standard data is formatted to ensure that data coming from one system's application layer can be recognised and read by another system.

The presentation layer provides services to the application layer. On the sender's side, the presentation layer transforms and formats the data passed down from the application layer and then sends the processed data to the next layer (the session layer). On the receiving side, the presentation layer is responsible for reverse processing the formatted data from the session layer, reducing it to the original data and then forwarding it to the application layer.

Data encryption, decryption, compression and decompression are all related to the presentation layer. The presentation layer of the sender encrypts and compresses the data passed down from the application layer as needed; the presentation layer of the receiver gets the data from the lower layer (session layer) and does the work of decompression if it is compressed, and decryption if it is encrypted, and finally sends the processed data to the application layer.

3. Session Layer

The session layer is responsible for establishing, managing and terminating sessions between presentation layer entities, and it coordinates and organises communication between systems. It provides a mechanism for clients' applications to open, close and manage sessions, i.e., semi-permanent dialogues. The session entity contains requests for session links to other programs and responses to requests for session links from other programs. In an application's operating environment, the session layer is the place where these programs make remote procedure calls (RPCs).

4. Transport Layer

The transport layer provides end-to-end communication services for application processes. It may include the following functionality:

Connection-oriented communication: it is often easier for an application process to interpret connections as data streams rather than dealing with the underlying connectionless model.

Same-order delivery: the network layer typically does not guarantee that packets arrive in the same order as they are sent. This is usually done by numbering the message segments and the receiver passes them on to the application process in order.

Reliability: Packets can be lost in transit due to network congestion and errors. With error checking codes (e. g., checksums), transport protocols can check for data corruption and confirm correct reception by passing an ACK or NACK message to the sender. An automatic retransmission request scheme can be used to retransmit the lost or corrupted data.

Flow control: Sometimes the data transfer rate between two nodes must be controlled to stop fast senders from transmitting more data than the receiving buffer can handle, causing a buffer overflow. This can also improve efficiency by reducing buffer underruns.

Congestion Avoidance: Congestion avoidance controls the amount of traffic entering a telecommunication network.

Multiplexing: Ports can provide multiple endpoints on a single node. For example, the name of a postal address is a form of multiplexing and distinguishes between different recipients at the same location. Each computer application process listens to their own port, which allows multiple network services to be used at the same time.

In order to complete the work of transport layer, two protocols have been designed, TCP and UDP. These two protocols work in parallel and are disjointed. But they accomplish the work defined for the transport layer differently: the TCP reliably ensures the reliable transmission of data through various policies, but the UDP ignores a lot of details and improves the efficiency of data transmission by sacrificing reliability. Programmers usually have to decide whether they want to use the TCP or the UDP for transmission, depending on the importance of the data to be transmitted.

5. Network Layer

The network layer sits between the data link layer and the transport layer and provides services to the transport layer. It encapsulates the data segments sent to it by the transport layer and generates IP packets; at the same time, it decapsulates the IP packets sent up from the data link layer and sends the data segments up to the transport layer.

The network layer mainly provides routing and addressing functions, so that the two terminal systems can be interconnected and determine the best path, and has a certain congestion control and traffic control capabilities. It is equivalent to the need to address the general importance of sending mails. As the network layer function in the TCP/IP system is specified and implemented by the IP, it is also known as the IP layer.

The network layer has two functions:

Addressing: The network layer uses IP addresses to uniquely identify devices on the Internet, and it relies on IP addresses to enable hosts to communicate with each other.

Routing: For different networks to communicate with each other, it is necessary to use routers and other Layer 3 devices, the network layer in the Layer 3 devices to create a routing table, in the routing table to save the best path to reach a certain network. When a packet needs to be transmitted, the network layer must make the decision to forward the data according to the routing table.

The main network devices that work at the network layer are: routers, layer 3 switches.

6. Data Link Layer

The data link layer provides management of the creation, maintenance, and release of data link connections between two network entities. It encapsulates the data link, data unit, the frame (frame), and controls frame delimitation, synchronisation, and send/receive order. The data link layer is also responsible for traffic control, error detection and error control during transmission.

The data link layer is located between the physical layer and the network layer, it provides services for the network layer, the network layer will send down the IP packet encapsulation into a data frame (frame), the frame format by the physical layer of the transmission medium. The format of the frame is determined by the transmission medium of the physical layer. For different transmission mediums, the format of the frame encapsulated by the data link layer is different.

In order to ensure the security of data transmission, it places a checksum (Parity, Sum, CRC) at the end of the frame to check whether the frame has been corrupted during transmission, thus transforming the potential error-prone physical connection provided by the physical layer into a logically error-free data link.

The data link layer of Ethernet is also responsible for controlling access to the medium. Access to the medium is controlled through Carrier Sense Multiple Access with Collision Detection (CSMA/CD) and Carrier Sense Multiple Access with Collision Avoidance (CSMA/Avoidance). Avoidance (CSMA/CA) mechanisms determine how to access the transmission medium in the conflict domain to improve data transmission efficiency. The data link layer is only responsible for data transfer from one end of the transmission medium to the other.

The main devices working in the link layer are: bridges, switches.

7. Physical Layer

The physical layer is the lowest and most fundamental layer in the OSI reference model of computer networks. Simply put, the physical layer of a network is responsible for sending and receiving data bits (bits, data bits) over a variety of physical medias.

The physical layer defines the electrical, mechanical, gauge, and functional requirements that need to be met in order to activate, maintain, and disconnect physical links between end systems. It specifies the physical layer connectors and various physical topologies in the form of standards that allow different systems to communicate with each other. For example, network ports, if the standard is not standardised, it is obvious that devices made by different manufacturers with ports and connectors of different sizes, or with different functions for each pin, are not compatible.

The main devices that work on the physical layer are hubs. However, due to its low working efficiency, it has been eliminated and will not be discussed here.

Exercise: Describe briefly in your own words the function of each layer of the OSI Network Reference Model.

Competence Expansion: The OSI reference model divides the entire task process of network data transmission into seven pieces, each piece of work is done by a separate layer, a total of seven layers, each layer has its own role. Each layer is indispensable in the communication process, but the work of a particular layer is meaningless if it is cut off from the whole model. Just as in society, people exist both as individuals and as members of a group, the group and individuals cannot be separated. The interests of the State, society as a whole and individuals are dialectically unified.

Task 2: Parsing a Simple Data Communication Process—Collaboration of OSI Layers

Mission Objective: Understand that the layers of the OSI reference model collaborate with each other to complete a basic data communication process.

How do the OSI layers work together to communicate data? Let's describe it through the simplest and most commonly-experienced example-the process of browsing the web:

Parsing a Simple Data Communication Process— Collaboration of OSI Layers

①The user opens Internet Explorer and enters the address of the server in the address bar.

②The IE browser calls the application layer HTTP to generate a request packet ready to be sent to the server to request the relevant web page.

③After the server receives the request from the client, it searches for the web page requested by the client and finally sends the web page data back to the client.

④After the client gets the web page data, it is displayed in Internet Explorer and the user can see the requested web page.

As shown in Figure 4.2, it is the whole process of the client sending the HTTP request to the server:

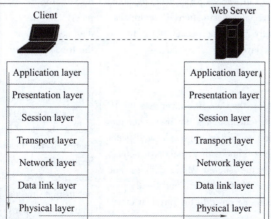

Figure 4.2 Client HTTP Request Sent to the Web Server

Table 4.1 describes the process by which a client sends an HTTP request to a Web server, based on the OSI reference model.

Table 4.1 Data Communication Process

Reference model	Client	Web server
Application Layer (computing)	(1) Generate HTTP request data, send the data down to the presentation layer	(14) After the HTTP service gets the request data, it looks for relevant resources according to the client's request
Presentation Layer	(2) Formatthe data, complete compression and encryption as required, and pass the processed data down to the session layer	(13) After the presentation layer gets the data, if the data is compressed, it needs to be decompressed; if the data is encrypted, it needs to be decrypted to get the most original data-HTTP request data. Finally, the data is sent upwards to the HTTP service in the application layer
Session Layer	(3) Start a session to pass data down to the transport layer	(12) Once youhave the data, send it up to the presentation layer
Transport Layer	(4) Give the data to the TCP for processing. The TCP is responsible for slicing the data and numbering each data slice; then the data slice is encapsulated into a data segment, and various parameters are filled in the header of the data segment, including the source port number and the destination port number; after the three handshakes are completed, the data segment is passed down to the network layer	(11) After getting the data segment, decapsulate the data segment to get the data slice, and then return an acknowledgement message to the sender's TCP to indicate that the data has been received; at the same time, reorganise the data according to the serial number in the header of the data segment to get the initially complete data, and send the data upwards according to the destination port number, to the session layer
Network Layer	(5) After the network layer gets the data segment, it encapsulates the data segment into an IP packet. Fill in the header of the IP packet with various parameters, including the source and destination IP addresses, and finally pass the packet down to the data link layer	(10) After getting the IP packet, check the destination IP address in the header of the IP packet. If you are not entitled to receive it, discard the IP packet; if you are entitled to receive it, decapsulate the IP packet, get the data segment, and send the data segment up to the transport layer
Data Link Layer	(6) After the data link layer gets the IP packet, it encapsulates the IP packet into a data frame. Fill in the header and footer of the data frame with various parameters, including the source MAC address and destination MAC address, and finally pass the data frame down to the physical layer	(9) Get the data frame from the physical layer, check the destination MAC address in the header of the data frame, if it is not entitled to accept the data frame, then discard the data frame; if the destination MAC address in the header of the data frame is your own or broadcast MAC address, then accept it and decapsulate the data frame to get the IP packet, complete the parity check and then upload the IP packet and send it to the network layer

continued

Reference model	Client	Web server
Physical Layer	(7) The physical layer converts data frames into a stream of binary bits and sends them out over the medium	(8) After receiving the binary bit stream, the binary bit stream is converted to a data frame, and the data frame is sent upwards to the previous layer

Based on the HTTP request from the client, the server finds the relevant resource and sends the resource data (which may be a web page, a picture or audio/video media, etc.) back to the client. The process is the same, except that the direction is just the opposite of the direction in which the request packet is sent.

The sender and the receiver do the opposite. The most important work of the sender is to encapsulate the data three times, and finally in the physical layer, the data will be converted into a binary bit stream sent through the network medium; correspondingly, the most important work of the receiver is to receive the binary bit stream from the network medium and convert it into frames, and then three times decapsulation, and finally get the original data generated by the sender to the application layer.

This process is like when we send a courier. We will put what we want to send into a box, and then put a sheet for this box, which is filled with a lot of information, including the sender's address and the receiver's address. The difference is that network communication is encapsulated three times, whereas we only need to encapsulate once to send a courier. For the receiver, he or she sees that it is his or her address, name, phone number and other information on the sheet of the courier, and confirms that it is his or her parcel before opening it. The receiver in the network communication should also do the corresponding work: after confirming that it is their own data corresponding to the data link layer, the network layer and the transmission layer for a total of three times to be decapsulated to get the most original data.

If we must take the network data forwarding and courier transmission to do an analogy, then according to the network data transmission way to transmit the courier, we should send the goods to be encapsulated in a small box, stick a list, and then put the small box into the middle of the box, stick the second list, and finally put the middle of the box into the big box, stick the third list, to complete the three times encapsulation process. When the receiver gets the courier, he has to unpack the big box first, then the middle box, and the small box, and finally gets the goods sent by the sender.

Why should encapsulate so many times? Simply put, to ensure that the data can be transmitted smoothly. For example, if someone buys a mobile phone online from abroad, the business is not to throw the phone directly to the ship and then through a long period of wind, rain and sunshine to the hands of the user. Rather, in order to protect the mobile phone, the manufacturer will first be encapsulated into the mobile phone box, and then the business will be loaded into the mobile phone box to the courier box, and finally on board the ship, and also the courier box into the container. Each encapsulation has his value and significance, which should be but carefully designed not be taken for granted or dispensable. Each layer of the reference model handles data differently and requires different encapsulations. Even at the same layer, different protocols encapsulate different units of data.

The Protocol Data Unit (PDU) generated by the three encapsulations of the OSI reference model have different names. The first encapsulation is completed by the transport layer, and the encapsulated PDU is called a data segment (Segment); the second encapsulation is completed by the network layer, and the encapsulated PDU is called a packet (Package); the third encapsulation is completed by the data link layer, and the encapsulated PDU is called a data frame (Frame).

Exercise: Describe a complete data communication process in your own words.

Competence Expansion: The OSI reference model has seven layers, each with its own responsibilities, while each layer collaborates with the upper and lower layers to complete each data communication. In addition to focusing on our own work, we also need to be deeply aware that the division of labour in various industries is becoming more and more detailed, and almost every task requires teamwork to complete. We need to correctly understand the dialectical unity of the team and the individuals: on the one hand, the individuals can not be separated from the team, the team unites each worker's wisdom and strength together to form the huge creativity; on the other hand, the team is composed of a number of people, if the enthusiasm of the individuals can not be mobilised, there will be no collective creativity.

Task 3: Switch Participation in Data Communications—Data Link Layer

Switch Participation in Data Communications— Data Link Layer

Mission Objective: Understand the process of switching.

A switch works at the data link layer of the OSI reference model and is responsible for processing data frames and is the most used device in computer networks. Because it is responsible for the second layer of the OSI reference model, it is sometimes customarily referred to as a Layer 2 device.

1. The Switch Relies on the MAC Address Table to Forward Data Frames

A switch is simply responsible for receiving data frames from one port and forwarding them out another. The question is: With so many ports, how does a switch decide from which port to forward a data frame?

Figure 4.3 shows the network topology of the R&D department. The MAC addresses of the Laptop2 and Laptop3 hosts are 0001.42d1.e839 and 0060.5c74.2c99, respectively.

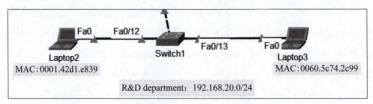

Figure 4.3　Network Topology of R&D Department

The switch's cache has a table that records host MAC address information (Mac Address Table), which can be viewed in the switch's privileged mode with the show mac-address-table command, as shown in Figure 4.4.

```
Switch1#show mac-address-table
          Mac Address Table
-------------------------------------------
Vlan    Mac Address        Type        Ports
----    -----------        --------    -----
 1      0001.c774.3002     DYNAMIC     Fa0/24
20      0001.42d1.e839     DYNAMIC     Fa0/12
20      0060.5c74.2c99     DYNAMIC     Fa0/13
```

Figure 4.4 MAC Address Table in the Switch

The meaning of the fields in the MAC address table is shown in Table 4.2.

Table 4.2 Meaning of Fields in the MAC Address Table

Serial number	Field	Significance	Note
1	Vlan	The VLAN in which the port is located	
2	Mac Address	MAC address of the host accessing the port	
3	Type	The way (type) of obtaining the MAC address of this host	DYNAMIC: Dynamic
4	Ports	port number	

If Laptop2 wants to send a data frame to Laptop3, the process is as follows:

①Laptop2 encapsulates the data frame with the source MAC address as its own MAC address and the destination MAC address as Laptop3's MAC address (0060.5c74.2c99).

②After encapsulation is complete, Laptop2 sends the data frame to the switch first.

③After the switch gets the data frame, it goes and reads the destination MAC address on the data frame.

④The switch checks the switch's cached MAC address table based on the destination MAC address on the data frame and determines that the host with MAC address 0060.5c74.2c99 is on port Fa0/13, so it forwards the data frame out of port Fa0/13.

⑤Laptop3 then gets the data frame from Laptop2 and the whole communication is complete.

This process is similar to the way a courier delivers a parcel. When the courier gets the parcel, he also has to look at the destination address on the parcel and then deliver it according to this destination address.

2. Learns the Source MAC Address of the Data Frame and Populates the MAC Address Table

For switches, the MAC address is so important that the forwarding of all data frames depends on this MAC address table. So where does this table come from? Is it there from the beginning?

No! The MAC address table in the switch is generated temporarily after the device boots up, saved in the cache, and is just an empty table at first. The records in the table are learned by the switch one by one when it forwards data in the following.

How does the switch learn MAC address records and keep populating the MAC address table?

When a switch gets a data frame from a port, it reads the source MAC address on the data frame to know which host is using the port and what its MAC address is. That MAC address and the port number of the corresponding switch are then added to the MAC address table to form an entry in the MAC address table. In the case of this task, at first the switch does not know which device is occupying port Fa0/12, let alone what the device's MAC address is. Until at some point, the switch receives a data frame from port Fa0/12, and then it goes to read the source MAC address on the data frame, and it learns that the MAC address of the device connected to port Fa0/12 is 0001.42d1.e839.

3. Flooding of Data Frames

What should be done if the switch cannot find the record of the destination MAC address in the MAC address table?

Since the MAC address table is empty at the beginning, if the switch gets a data frame and checks the MAC address table according to the destination MAC address on the data frame, it will not be able to find the corresponding record of the destination MAC address, and therefore it will not be able to determine which port the target host is on. At this point, how will the switch dispose of the received data frame?

The answer is: flooding. Since we don't know which port the destination host is on, we can use a similar method of "broadcasting" to make many copies of the data frame and send one copy to each port, so there is always a port that is correct. Even if none of the ports is correct, the loss of resources is small.

This type of flooding is a last resort, as it poses a number of security problems. Firstly, it takes up the switch's CPU and cache resources, but this is not the main one. What is more risky is that the act of flooding sends data frames to every port, which can give sniffers in the network an opportunity to illegally access the data, thus leading to information leakage. At the same time, hackers may use tools to forge invalid data frames with different MAC addresses and keep sending them to the switch, which learns new records and writes them to the MAC address table until the MAC address table is filled. When the real user data is sent, the switch learns a new valid MAC address record, but it is no longer able to write to the MAC address table which is already full. Afterwards, if a data frame is to be sent to a host with that MAC address, the switch cannot find a record for the destination MAC address in the MAC address table. So the data frame can only be forwarded by flooding, we can solve this problem with techniques related to switch port security.

Exercise: Describe in your own words how the switch works.

Competence Expansion: The switch makes the decision to forward data frames based on the MAC address table; when a record of the destination MAC address is found in the MAC address table, it forwards the frame out according to the port of the record; if no record of the destination

MAC address can be found in the MAC address table, the data frame is sent out by means of flooding. If the security factor is ignored, then the flooding method reduces the possibility of data frame loss as much as possible. In our studies and future work, we will inevitably face a variety of difficulties and setbacks, to do as much as possible, to try and not give up. Setting up a scientific and lofty ideals and beliefs will provide us with the impetus to move forward, will increase our courage in the face of setbacks and confidence in overcoming difficulties.

Task 4: Router Participation in Data Communications—Network Layer

Router Participation in Data Communications— Network Layer

Mission Objective: Understand how a router (Layer 3 switch) works.

Routers (Layer 3 switches) work at the network layer of the OSI reference model, are responsible for processing IP packets, and are among the most used devices in computer networks. Because it is responsible for the third layer of the OSI reference model, it is sometimes habitually referred to as a Layer 3 device.

1. The Process of Router Forwarding IP Packets

As shown in Figure 4.5, R0 is a border router for the company's network, connected to the company's intranet on the left and the carrier's network on the right.

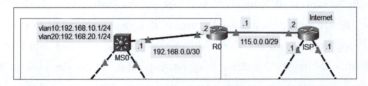

Figure 4.5 Corporate Network Access to ISP Network Topology

The router is responsible for a very simple job, which is to receive IP packets from one port and forward them out from another. The question is: how does a router with many ports know which port to forward the IP packet before it delivers the IP packet to the receiver?

The answer is: to check the routing table. There is also a table in the router's cache, which can be viewed in the privileged mode with the show ip route command, as shown in Figure 4.6.

When the router receives an IP packet, it will read the destination IP address on the IP packet, and then check the routing table according to the obtained destination IP address, and find the corresponding routing record according to the principle of the longest match of the subnet mask, and then finally decide to forward it out through the port specified in the routing record.

For example, if the router receives an IP packet and finds that the destination address is 192.168.0.1, it goes to check its routing table according to this address, matches the second record in the table-the routing entry of 192.168.0.0, and then forwards the IP packet from the port FastEthernet0/0 and then forwards the IP packet out of the FastEthernet0/0 port. This process is

```
R0#sh ip route
Codes: C - connected, S - static, I - IGRP, R - RIP, M - mobile, B - BGP
       D - EIGRP, EX - EIGRP external, O - OSPF, IA - OSPF inter area
       N1 - OSPF NSSA external type 1, N2 - OSPF NSSA external type 2
       E1 - OSPF external type 1, E2 - OSPF external type 2, E - EGP
       i - IS-IS, L1 - IS-IS level-1, L2 - IS-IS level-2, ia - IS-IS inter area
       * - candidate default, U - per-user static route, o - ODR
       P - periodic downloaded static route

Gateway of last resort is 115.0.0.2 to network 0.0.0.0

     115.0.0.0/29 is subnetted, 1 subnets
C       115.0.0.0 is directly connected, FastEthernet1/0
     192.168.0.0/30 is subnetted, 1 subnets
C       192.168.0.0 is directly connected, FastEthernet0/0
S    192.168.10.0/24 [1/0] via 192.168.0.1
S    192.168.20.0/24 [1/0] via 192.168.0.1
S*   0.0.0.0/0 [1/0] via 115.0.0.2
```

Figure 4.6　Routing Table for Router R0

also similar to that of a courier delivering a parcel in that it is based on the destination address on the parcel that determines how to deliver it.

It is worth noting that after the router forwards the packet, the task is over, and it does not care whether the other end can receive it. It doesn't even care whether the destination host exists. As long as it matches the appropriate route entry in the routing table, it forwards the packet according to the port number marked by the route entry.

2. Router Drops IP Packets

What if I can't match the corresponding route entry in the routing table?

For a router to forward an IP packet, it will look up the routing table based on the destination IP address on the packet, and discard the IP packet if it does not match the corresponding route entry. This is different from a switch, which will use flooding to send the data frame to each port if it cannot find an entry in the MAC address table corresponding to the destination MAC address.

But there is one detail that seems particularly gentlemanl after a router drops an IP packet. The router sends an ICMP (Internet Control Message Protocol) packet to the sender, telling the sender: Because I can't match a route entry here, your IP packet has been dropped by me!

3. Details of IP Packets to and From the Router

It is customary to say that a router works at the network layer of the OSI reference model, but in fact it has to do the work of all three layers including the network layer below. The process of packets moving in and out of the router is shown in Figure 4.7.

The entire process of processing data by the router is as follows:

①The router receives the binary bit stream from the physical layer and converts the resulting binary bit stream into a data frame.

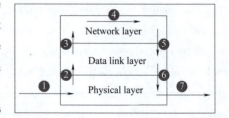

Figure 4.7　Processing of Packets moving in and out of the Router

②The physical layer of the router transmits the data frame upwards to the data link layer.

③When the data link layer of the router gets the data frame, it reads the destination MAC address on the data frame and determines whether it has the right to accept it or not (if the MAC address is its own, or if this destination MAC address is a Layer 2 broadcast address, then it accepts the frame), and then it decapsulates the data frame to get the IP packet, and finally the data link layer uploads the IP packet towards the network layer.

④After the network layer of the router gets the IP packet, it reads the destination IP address on the packet, checks the routing table record based on this IP address, and discards the packet if it does not match.

⑤If a match is made, a decision is made as to which port to send the packet out of based on the matching route entry. But the port works in the physical layer and sends the IP packet down to the data link layer.

⑥After the data link layer of the router gets the IP packet, it encapsulates the IP packet into a new data frame and writes the new source MAC address (the MAC address of its own outgoing port) and the new destination MAC address (the MAC address of the next-hop node) to the data frame, and finally sends the new data frame down to the physical layer.

⑦The physical layer of the router gets the data frame and converts it into a binary bit stream which is forwarded out through the network port to the next node.

Routers are used to connect different networks and broadcast packets are restricted to be transmitted within the network, so routers are good at isolating broadcast traffic.

Exercise: Describe in your own words how the router works.

Competence Expansion: A router decides how to forward data based on the routing entries in the routing table: it forwards when it matches a routing entry and discards when it doesn't. Packets are dropped to prevent some packets from taking up too many network resources, which can affect normal network communication. Sometimes, giving up is not always a bad thing. We should not stick to the immediate gains and losses, we should set up a correct view of gains and losses, correctly understand and deal with the development process of life in the of this contradiction gains and losses, to achieve the value of life.

Task 5: TCP Ensures Reliable Data Transfer—Transport Layer

Mission Objective: Understand how the TCP ensures the reliable transmission of data.

To explore the transport layer of the OSI reference model in depth, it is important to observe where the transport layer is located in the OSI reference model, as shown in Figure 4.8.

The transport layer is between the session layer and the network layer,

TCP Ensures Reliable Data Transfer— Transport Layer

providing services to the session layer and relying on the network layer for data transfer. When a host has data to send, the data travels from the application layer all the way down to the transport layer. The transport layer is responsible for slicing and dicing large pieces of data, encapsulating the pieces of data into segments, and populating the header of the segment with information, including the source and destination port numbers.

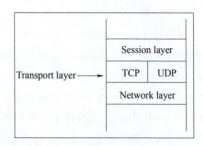

Figure 4.8　Location of the Transport Layer in the OSI Reference Model

What is the port number? A port number is not the number of a physical port on a device, but is an abstraction of the number assigned to a system service or an application when it participates in network communication. In other words, port numbers identify different applications or services on a host. The source port number on a data segment identifies which application is sending the data from the sender, while the destination port number identifies which application is sending the data to the receiver.

The transport layer uses port numbers to mark the applications or services of the application layer that are involved in network communication. The sender's transport layer is responsible for marking which application of the host the data is sent from, and the receiver's transport layer gets the data segment from the network layer and sends the data to the appropriate application based on the destination port number. As to which host one belongs and to which host the data is ready to be sent, this task is left to the network layer, the transport layer doesn't care.

According to the results of the analysis of Task 2 of this project, the main responsibility of the transport layer on both sides of the data communication is different. The main responsibilities of the transmission layer of the sender include: slicing the data, encapsulating the data slices into data segments, and filling in various parameters in the header of the data segments, including the source port number and the destination port number. The main responsibilities of the transmission layer of the receiver include: de-encapsulating the obtained data segments to obtain the data slices, reassembling the data slices, obtaining the initial complete data and sending the data to the corresponding service or application according to the destination port number sent to the corresponding service or application.

In order to complete the work of the transport layer specified by the OSI reference model, we have designed many protocols at the transport layer, but the main ones are the TCP (Transmission Control Protocol) and the UDP (User Datagram Protocol). Both protocols can independently complete the tasks specified in the transport layer, and the TCP and UDP do not need to collaborate with each other.

The main difference between the TCP and UDP is the quality of the data transmission provided; the TCP is connection oriented, which ensures the reliability of the data transmission, while the UDP is non-connection oriented, which provides an unreliable, best-effort data transmission service for the upper layers, and does not guarantee that the data will be sent to the receiving party without any problems. Simply put, if the sender gives the data to the TCP for transmission, the sender's TCP will be synchronised with the receiver's TCP beforehand (connection-oriented), and the data will be transmitted only after agreement is reached. However, if the sender gives the data to the transport layer UDP, the situation is completely different. The sender's UDP will work hard to send the data, but it will not try to synchronise with the receiver's UDP beforehand (non-connection oriented), it doesn't care if the receiver is ready to receive the data, or even if the receiver is alive in the network, it doesn't care. As long as the upper layer has data to send, it just sends it without guaranteeing that the data will reach the other party, and it is obvious that such a transmission method is unreliable.

To analyse the TCP specifically, it mainly takes the following four measures to ensure the reliable transmission of data between the two sides of the communication.

1. Connection-oriented (three handshakes)

Connection-oriented, that is, before the data transmission, the sender and the receiver send messages to each other to achieve synchronisation and to make sure that both the sender and the receiver are ready. The process is shown in Figure 4.9.

The TCP sends and receives the two sides to obtain communication before the synchronisation They need to send each other their own synchronisation messages (Synchronize Sequence Numbers, SYN) to the other side, at the same time in the receipt of the other side of the message to return to the other side of the confirmation (Acknowledge, ACK) message. This process requires both parties to send each other a total of three messages, known as the TCP protocol three handshakes.

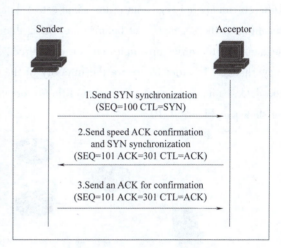

Figure 4.9 TCP Three Times Handshake Process

The first handshake: initiated by the sender, a synchronisation message SYN to the receiver is sent, indicating that it wants to send data to the receiver to remind him of getting ready. The "Serial Number" SEQ value of the synchronisation message is a random number, in this case 100.

The second handshake: the receiver receives the synchronisation message from the sender, according to the rules must give the sender an acknowledgement ACK (the value of the ACK is the SEQ value of the received SYN plus 1, here the first handshake was sent when the SEQ value of 100, in order to indicate that it has received this SYN synchronisation, the receiver will be the value of the value of the SEQ value plus 1 value of 101 to send back to the sender), indicating that the message has been received. It also sends its own SYN synchronisation message to the sender, and the SEQ value sent at this time is also random, in this case 200.

The third handshake: the sender receives the synchronised SYN message of the second handshake from the receiver, and according to the rules, it has to give an acknowledgement, and the value of the ACK is also equal to the SEQ value of the second handshake plus 1 (here it is 301); at this time, the value of SEQ is no longer random, but is the value of the ACK of the second handshake.

Once the three handshakes are complete, the sender and receiver are synchronised (ready), i. e. the receiver is ready to receive and the sender is ready to send, and communication can begin.

2. Data Slicing, Reorganisation Mechanisms

The first thing the sender's transport layer does when it receives data from the upper layers is to slice the user's data and encapsulate the slices to obtain data segments. These data segments are then numbered and the number is placed in the header of the data segment.

The role of the sender to number the data is to facilitate the receiver to receive a set of data segments, the reorganisation work is to be done. It ensures that the data is reassembled back in the original order and that there are no errors in the data.

In an intricate network, the transmission of data does not always arrive in the desired manner. Different numbered segments of the same set of data may travel different paths during transmission, resulting in the segment sent before being later than the one sent later. The sender sends the data segments in the order in which they are numbered, but the order in which they arrive at the sender may change. In order to ensure the integrity of the data, the receiver's TCP protocol will first sort the data segments by their number sand then reorganise them to get the original data without errors, as shown in Figure 4. 10.

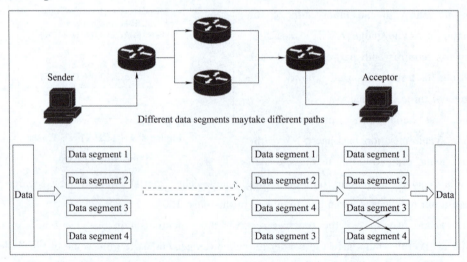

Figure 4. 10　The Receiver's TCP Protocol Reorganises the Data Segments After Ordering Them

3. Flow Control Mechanisms for Sliding Windows

The TCP also provides a flow control mechanism, i. e. , the amount of data that can be reliably accepted and processed by the receiver. The flow control helps to maintain the reliability of TCP transmissions by adjusting the rate of data flow between the source and the destination for a given session. For this purpose, the TCP header includes a 16-bit field called the "window size".

The initial window size is determined when the TCP session is established during the three handshakes. The sender must limit the number of bytes sent to the receiver based on the window size of the destination device. The sender can continue to send additional data segments only after it receives an acknowledgement (ACK) that the data segment has been received. Typically, the receiver does not wait for all bytes in its window size to be received before responding with an acknowledgement. As bytes are received and processed, the receiver sends an acknowledgement to inform the sender that it can continue to send more bytes.

The receiver processes the received data segments and continually adjusts the size of the acknowledgement window back to the sender, a process known as a sliding window. Typically, the receiver sends an acknowledgement after receiving every two data segments. The number of data segments received before the acknowledgement may vary. The advantage of a sliding window is that it allows the sender to keep transmitting data segments as long as the receiver acknowledges the previous data segments.

The receiver generally adjusts the window size based on the size of its own cache and the loss rate of data segments. If its own cache is already very small, the sender will adjust the window to a smaller size and tell the sender to reduce the amount of data segments sent, so as not to cause the receiver to discard the received data segments due to insufficient cache on the contrary, if the receiver still has a large cache, it will adjust the window to a larger size and then notify the sender; if the receiver receives data segments with a very high rate of loss, it is judged that it may be due to the transmission of network blocking, network equipment to avoid congestion, discard part of the data. At this time the receiver will also reduce the window, tell the sender to reduce the number of data segments sent to reduce the transmission channel on the flow, to alleviate the blockage of the network path to improve the success rate of data transmission, and vice versa, if the receiver finds that the loss of data rate is very low, it will be adjusted to a larger window, and then notify the sender of the size of the window.

4. Acknowledgement, Retransmission Mechanisms

According to the rules of TCP data transmission, in the receipt of data sent by the sender, the receiver gives the sender an acknowledgement (ACK), in such a way as to tell the sender, sent data segments have been received This is the confirmation mechanism.

The acknowledgement message from the sender to the receiver (the value of the ACK) has another meaning: please send me the data segment with the serial number equal to the value of the ACK. This is shown in Figure 4.11.

Due to the complexity of the network and the influence of the environment, it is inevitable that there will be errors or loss of data in the process of transmission Confirmation and retransmission mechanism, to a large extent, are to ensure the integrity of the data transmission process.

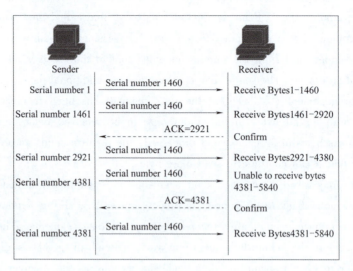

Figure 4.11　TCP Protocol Acknowledgement, Retransmission Mechanism

Exercise: Describe in your own words the similarities and differences in the way the TCP and UDP work in the transport layer of the OSI Reference Model.

Competence Expansion: The TCP, in order to ensure the reliable transmission of data, mainly adopts the following mechanisms three handshakes, slicing and reorganisation of data, sliding-window flow control, acknowledgement of retransmission, etc. The TCP is a reliable and responsible protocol. We also need to develop our own sense of responsibility. Learn to be responsible for yourself, your loved ones, the people around you and the nation, the state and society, and try to be a valuable and responsible person. To correctly understand and deal with the various problems encountered in life, can not get by, indulge life, game life, otherwise it will waste time, or even go astray.

Summary and Expansion

1. Comparing the OSI Reference Model and the TCP/IP Model

A network model is an abstract representation of a computer network. The computer network model published by the ISO organisation is called the OSI reference model, which is divided into seven layers. But there is a more concise model, also recognised and accepted by the industry, called the TCP/IP model. The correspondence between these two models is shown in Figure 4.12.

The role of a computer network is to enable data transmission, and a network model is just an abstract description of a computer network. There are many ways to describe it, and many network models can be proposed, but no matter how to describe it, the function of computer networks is the same. In other words, no matter what model is used to describe a computer network, the functions of the network defined by these models are the same.

Project 4 Analyzing Network Communications

The TCP/IP model and the OSI reference model are just different descriptions of a computer network, and these descriptions do not affect the essence of a computer network-the realisation of data transmission.

The difference is that the TCP/IP model describes the computer network in a more concise way, the TCP/IP model unifies the functions of the upper three layers of the OSI reference model (the session layer, the presentation layer, and the application layer) into a single layer, i.e., the "Application layer", and the functions of the lower two layers (the Physical layer, and the Data Link layer) into one layer, i.e., the "Network Access layer". The next two layers (physical layer and data link layer) are unified into one layer, i.e. "Network Access layer", the middle transport layer still corresponds to the transport layer, and the network layer also corresponds to the other layer, but with a different name which is called the network interconnection layer (Internet Layer).

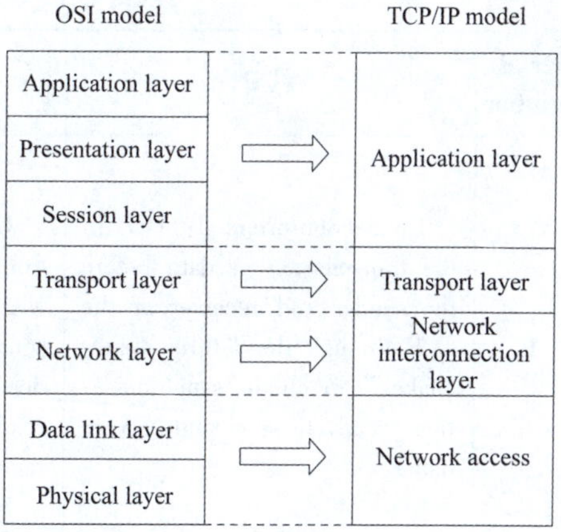

Figure 4.12 Correspondence between the OSI Reference Model and the TCP/IP Model

Just because the OSI reference model describes a computer network in 7 layers and the TCP/IP model describes a computer network in 4 layers, it does not mean that the TCP/IP model has less work to accomplish. In fact, the amount of work they have to accomplish is the same, only the work has been reorganised. It's like a company reorganizes its business units, but no matter how they are put together, the total amount of work is still not reduced or increased for the company to function properly.

2. Comparison of TCP and UDP

There are two main protocols in the transport layer, one is the TCP and the other is the UDP. These two protocols are parallel and disjointed, each of them independently does the work specified in the transport layer. So what is the difference between them? If they can each do their jobs independently, why should two protocols at the transport layer be designed instead of developing just one?

The difference is that the TCP is a reliable protocol and the UDP is an unreliable protocol. When transmitting important data, we will more often consider giving the data to the TCP to let it be responsible for the transmission, while for some data that is not very important and has high latency requirements, we can consider giving it to the UDP to be transmitted.

The reason why the TCP is reliable has been analysed earlier, so to what extent is the UDP unreliable compared to the TCP? Let's compare the two protocols through a Table 4.3.

Table 4.3 Comparison of TCP and UDP

Serial Number	TCP	UDP	Note
1	Connection-oriented: TCP in the transmission of data before the sender and receiver of the TCP through the "three handshakes" to obtain synchronisation, ready to send and receive data	Oriented towards no connection: The UDP sends data without synchronisation. The sender's UDP with just send, it doesn't care if the receiver is ready to receive or even if the receiver exists	The TCP establishes virtual circuits on both sides of the sender and receiver through "three handshakes", which takes up time and other overheads; the UDP doesn't require any overheads until the data transmission starts
2	Confirm retransmission: After the receiver's TCP receives the data, it will give an acknowledgement message to the sender, indicating that the data segment has been received; if the data is corrupted or lost during transmission, the receiver will also ask the sender to send the data segment with the corresponding serial number again through an ACK message	No confirmation: The receiver's UDP does not send an acknowledgement message to the sender's UDP after receiving the data segment. The sender and receiver are like two sulking children; you don't tell me when you send, and I don't want to tell you when I receive	The TCP ensures that data reaches the opposite end without error by confirming retransmission, which increases the workload and affects the efficiency of data transmission; the UDP does not have any measures, and data transmission may be erroneous, but it is very efficient
3	Sort of reorganisation: The sender's TCP numbers the different data segments when slicing the data; when the data segments arrive at the destination, the receiver's TCP reorganises them according to the numbering of the data segments to ensure the integrity of the data	Disorderly reorganisation: The sender's UDP is not numbered when it slices the data. After encapsulating the data segments, they are sent out; obviously, most of the time, the order in which the data arrives at the receiver is messed up, and at this point, the receiver's UDP reorganises the data in the order in which it arrives, and because the sender has not numbered the data segments, they cannot reorganise them by number, which can result in messed up data	The receiver's TCP must reserve more cache to store all the data segments of the same set of data, and must wait until all the data segments have arrived before they can be reorganised in a shot; if the UDP is used for transmission, the receiver is reorganised on a first-come, first-served basis, which doesn't require too much cache, and at the same time, work efficiency is improved

continued

Serial Number	TCP	UDP	Note
4	Sliding window flow control: The two senders of the TCP determine the size of the window during the third handshake, and the receiver will adjust the window size according to its own cache size and the loss rate of the data segment and notify the sender. The sender will send data according to the window size to alleviate possible network blocking and ensure the success rate of data transmission	There is no flow control mechanism: The UDP endeavours (with the maximum possible amount of data) to send data when there is data to be sent, regardless of whether the network appears to be blocked, without any control, potentially leading to an avalanche effect by causing the network to become more congested	
5	Newspaper header size: 20 Byte	Newspaper header size: 8 Byte	A larger TCP header than a UDP header results in higher overhead

Comparison from the table can be seen, the UDP is less reliable, but the overhead is small, the efficiency of the transmission of data than the TCP is higher. The TCP can indeed guarantee the reliability of data transmission, but also pay a lot of resource overhead, affecting the efficiency of data transmission. In fact, before the computer network equipment manufacturing process is not very fine when the network equipment to deal with data will indeed bring some data loss phenomenon, but with the improvement of the technical level of the industrial process, the network data transmission quality relative to the birth of the computer network has been a great improvement in the early days so, very often, we can not have to think about too many factors, such as resorting to the TCP to guarantee the reliable transmission of data.

3. Difference Between the Two Ways in Which Switches Forward Data: Broadcasting and Flooding

A switch broadcasts or floods data frames, and the behaviour is the same: N copies of the data frame are made and one copy is sent out of per active port. The difference is that the destination MAC address of a broadcast frame is a broadcast address, and the destination MAC address of a flooded frame is the address of the destination host, which is the MAC address of a specific particular device.

Ethernet is a broadcast-type network. When the data link layer encapsulates an Ethernet frame, it writes the source MAC address and the destination MAC address into the header of the frame. The source MAC address is the MAC address of the host sending the data itself, which can be determined.

But when the sender wants to send a frame to all the hosts in the network, it writes a Layer 2 broadcast address (full F) in the destination MAC address field in the header of the data frame, and the behaviour of the switch in forwarding such frames is called broadcasting.

When a switch receives a data frame, it then looks up the MAC address table based on the destination MAC address on this data frame, but if it cannot find it in the table, it sends out the data frame by using flooding.

4. ARP queries

Let's start by analysing the situation in which the sender and the receiver are on the same network. For example, the sender wants to send a set of data to host 192.168.0.1.

On the sender's side, when the data goes all the way down to the data link layer ready to be encapsulated a third time, the source MAC address and the destination MAC address need to be filled in the header of the data frame. The source MAC address is its own address, which the sender knows. But when it comes to filling in the destination MAC address, the sender does not know the MAC address of the destination host.

In order to complete the encapsulation of the frame, the sender must ask the receiver for the MAC address, but it does not know which device is the destination host, so it sends out a broadcast frame (the destination address of the frame is the full F), and this broadcast frame is the ARP query. This broadcast frame is an ARP query, which reads: Who is 192.168.0.1, please send me your MAC address. When the destination host receives the request, it returns an answer, telling the sender its MAC address. After the sender gets the MAC address of the destination host, the encapsulation of the data link layer frame continues.

Because ARP queries are sent out as broadcasts, direct ARP queries between the sender and the receiver are not possible if the sender and the receiver are not on the same network, because ARP broadcast frames cannot be transmitted across networks. Data frames are only responsible for transmitting data within a network. In a cross-network environment, data is transmitted down the network hop by hop. Each time it passes through a network, the data link layer must encapsulate a new frame in order to send the data frame to the next hop device. This is shown in Figure 4.13.

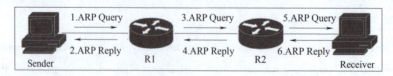

Figure 4.13　ARP Query between Different Networks

If the sender and the receiver are not on the same network, the ARP resolution proceeds as follows:

①The sender then first has to send the data to the gateway (R1).

②R1 returns its MAC address to the sender, who sends the data to R1 after completing the encapsulation of the data frame.

③After R1 gets the data frame, it decapsulates the data frame and sends the IP packet to the network layer for routing query. After R1 finishes the routing query, it decides that it wants to forward the data to R2; so R1 asks for the MAC address of R2 through an ARP query.

④R2 returns its MAC address to R1. The data link layer of R1 encapsulates the new data frame based on the MAC address of R2 obtained from the query, and then sends the new data frame to R2.

⑤R2 sends an ARP query to the receiver.

⑥The receiver returns its MAC address to R2, which encapsulates the new frame and sends it to the receiver to complete the last hop of data transmission.

5. Recursive Queries for Routers

Recursion, in mathematics and computer science, is the use of a function's own method in the definition of a function. In the case of routing table lookup, it refers to multiple lookups of the routing table, i. e., a way of going to a second lookup based on the results obtained from a previous lookup of the routing table.

The routing table for router R0 is shown in Figure 4.14.

```
R0#sh ip route
Codes: C - connected, S - static, I - IGRP, R - RIP, M - mobile, B - BGP
       D - EIGRP, EX - EIGRP external, O - OSPF, IA - OSPF inter area
       N1 - OSPF NSSA external type 1, N2 - OSPF NSSA external type 2
       E1 - OSPF external type 1, E2 - OSPF external type 2, E - EGP
       i - IS-IS, L1 - IS-IS level-1, L2 - IS-IS level-2, ia - IS-IS inter area
       * - candidate default, U - per-user static route, o - ODR
       P - periodic downloaded static route

Gateway of last resort is 115.0.0.2 to network 0.0.0.0

     115.0.0.0/29 is subnetted, 1 subnets
C       115.0.0.0 is directly connected, FastEthernet1/0
     192.168.0.0/30 is subnetted, 1 subnets
C       192.168.0.0 is directly connected, FastEthernet0/0
S    192.168.10.0/24 [1/0] via 192.168.0.1
S    192.168.20.0/24 [1/0] via 192.168.0.1
S*   0.0.0.0/0 [1/0] via 115.0.0.2
```

Figure 4.14 Routing Table of a Router

If R0 receives an IP packet with a destination network of 192.168.10.0, what is its process for querying the routing table?

First R0 will match to the entry in the routing table based on the destination address on the IP packet:

    S 192.168.10.0/24[1/0] via 192.168.0.1

It is decided that the packet is to be forwarded to the next hop, 192.168.0.1, but it is not known which port of R0 this device is connected to. Then a second query is made to the router to match to another entry in the routing table:

    C 192.168.0.0/24 is directly connected, FastEtherner0/0

From this route entry, it is known that the network where host 192.168.0.1 is located is connected to port FastEtherner0/0, and it is only at last that it is decided that the IP packet should be sent out from port FastEtherner0/0.

Obviously, in order to forward the IP packet out, the routing table is queried twice here, and the later query is based on the result of the earlier query.

6. IP Packet Life Cycle (Time To Live, TTL)

The TTL value is a field (8 bit) in the IP packet header used to limit the life cycle of the packet.

The initial TTL of an IP packet is set by the source device of the packet. When the packet is processed once by a router (Layer 3 device), the TTL value is decremented by 1. If the value of the TTL field is decremented to 0, the Layer 3 device discards the packet and sends an ICMP timeout message to the source device. Tells the sender of the IP packet that the packet has reached the end of its life cycle, but the destination host has not been found and the packet has been dropped.

7. Variable Length Subnet Mask (VLSM)

Let's start by reviewing IP addresses. We already know that IP addresses are used to identify different hosts on the Internet. A subnet mask is used to determine which bits in the IP address are host bits and which bits are network bits. The number of bits in the host bits determines the number of different IP addresses that can be expressed in that network. Assuming that the host bit of a particular IP address is n bits, it is easy to figure out that there are at most 2n different IP addresses in that network based on mathematical combinations. Of these IP addresses, the smallest address is called the network address, which is used to identify different networks, and the largest address is called the broadcast address, which is used to encapsulate broadcast packets. The remaining 2n-2 addresses can be assigned to hosts within the network for communication.

When a network has a large number of hosts and not enough IP addresses, the IP address expansion can be achieved by allowing the host bits to borrow bits from the network bits to increase the number of host bits. Similarly, if not so many IP addresses are needed and you do not want to result in wasted IP addresses, you can borrow subnet bits from the host bits to reduce the IP address capacity of the network.

Learn more about fixed-length subnet masks. If a company applies for a class C address: 211.1.1.0/24, and the company has 4 departments, it needs to be divided into 4 subnets. So the host bit can borrow 2 bits out as subnet bits, at which point the total length of the network bits is no longer 24, but 26, as shown in Figure 4.15.

As a result of the subnetting, each department has a space of 64 IP addresses, minus the network and broadcast addresses, leaving 62 IP addresses available for assignment to hosts. In other words, each department can accommodate up to

DepartmentA: 11010011.00000001.0000001.	00	000000	/26
DepartmentB: 11010011.00000001.0000001.	01	000000	/26
DepartmentC: 11010011.00000001.0000001.	10	000000	/26
DepartmentD: 11010011.00000001.0000001.	11	000000	/26
	Network bit	Subnet bit	host bit

Figure 4.15 Segmenting the Original IP Address Space

62 computers (including gateways).

However, if the number of people in each department is taken into account, then subnetting is done by using a fixed-length subnet mask, which results in wasted IP addresses as shown in Table 4.4.

Table 4.4 Statistics on the Number of IP Addresses by Sector

Sectoral	Computers (units)	Number of host addresses assigned	Number of wasted addresses
A	2	62	60
B	28	62	34
C	10	62	52
D	50	62	12

Obviously, a lot of IP addresses are wasted for the A, B, C divisions. Can the division of IP addresses be further optimised to save more IP addresses for future network expansion?

The waste of IP addresses can be minimised by using variable-length subnet masks. The idea and calculation steps are shown in Table 4.5.

Table 4.5 Ideas and Steps for Subnetting with Variable-length Subnet Masks

Move	Sports Event	Sector A	Sector B	Sector C	Sector D	Note
1	Number of computers	2	28	10	50	Determine the number of computers
2	Number of host addresses required	2	28	10	50	Each computer requires 1 IP address
3	Number of IP addresses required	4	30	12	52	Host address plus 2 special addresses (network address and broadcast address)
4	Number of IP addresses assigned	$2^2 = 4$	$2^5 = 32$	$2^4 = 16$	$2^6 = 64$	Cannot be assigned arbitrarily. It can only be allocated by 2n, whichever just meets the need. n is the number of host bits
5	Range of IP Addresses per subnet	211.1.1.112 – 211.1.1.115	211.1.1.64 – 211.1.1.95	211.1.1.96 – 211.1.1.111	211.1.1.0 – 211.1.1.63	The IP addresses are allocated according to the highest to lowest IP address demand, here in the order of D-B-C-A

continued

Move	Sports Event	Sector A	Sector B	Sector C	Sector D	Note
6	Subnet mask length	/30	/27	/28	/26	The subnet mask length is: 32-n
7	Network address	211.1.1.112/30	211.1.1.64/27	211.1.1.96/28	211.1.1.0/26	Smallest IP address in the subnet
8	Broadcast address	211.1.1.115/30	211.1.1.95/27	211.1.1.111/28	211.1.1.63/26	Largest IP address in the subnet
9	Range of Available host addresses	211.1.1.112/30 - 211.1.1.115/30	211.1.1.64/27 - 211.1.1.95/27	211.1.1.9/28 - 211.1.1.111/28	211.1.1.0/26 - 211.1.1.63/26	Intermediate range of IP address segments in the subnet

Note: When allocating IP addresses, they should be distributed to large networks before small networks, otherwise IP address fragmentation will occur, which is not conducive to future network expansion.

8. Port Number Basics

Port numbers at the transport layer are used to distinguish between different applications or services on the same host. The Internet Assigned Numbers Authority (IANA) is the standard body responsible for assigning various addressing standards, including port numbers. There are different types of port numbers as follows:

Recognised ports (ports 0 to 1023), these numbers are used for services and applications; applications such as Web browsers, E-mail clients, and remote access clients typically use these port numbers. By defining recognised ports for server applications, client applications can be set up to request connections to specific ports and their associated services.

Registered ports (ports 1024 through 49151), these port numbers are assigned by IANA to requesting entities for use by specific processes or applications. These processes are primarily some of the applications that the user chooses to install, rather than commonly-used applications that have been assigned with recognised port numbers. For example, Cisco has registered port 1985 as its Hot Standby Router Protocol (HSRP) process.

Dynamic or private ports (ports 49152 through 65535), also known as temporary ports, are typically dynamically assigned by the client operating system when a host initiates a connection to a server. Dynamic ports are used to identify client applications during communication.

Port Numbers for Common Applications or Services are shown in Table 4.6.

Table 4.6 Port Numbers for Common Applications or Services

Port Number	Transport Layer Protocol	Applications/Services	Abbreviations
20	TCP	File Transfer Protocol (data)	FTP
21	TCP	File transfer protocol (control)	FTP

continued

Port Number	Transport Layer Protocol	Applications/Services	Abbreviations
22	TCP	Safety enclosure	SSH
23	TCP	Telnet	-
25	TCP	Simple Mail Transfer Protocol	SMTP
53	TCP, UDP	domain name service	DNS
67	UDP	Dynamic Host Configuration Protocol (server)	DHCP
68	UDP	Dynamic Host Configuration Protocol (client)	DHCP
69	UDP	Simple File Transfer Protocol	TFTP
80	TCP	Hypertext transfer protocol (HTTP)	HTTP
110	TCP	Post Office Agreement Version 3	POP3
143	TCP	Internet Message Access Protocol (IMAP)	IMAP
161	UDP	SNMP	SNMP
443	TCP	Secure Hypertext Transfer Protocol (HTTP)	HTTPS

It is important to know the port numbers of some common applications or services, firstly because network management work is often involved, and secondly when designing network access policies to avoid affecting normal applications or services.

9. Sockets

In order to clearly express both sides of a data communication, we have to resort to both the concepts of the IP address and the port number.

A distinction is made here between IP addresses, which marks different hosts in a network, and port numbers, which mark different applications or services within a host that are involved in network communication. The source IP address identifies the host from which the data is being sent, while the destination IP address identifies the host in the network to which the data is to be sent. The source port number marks which application on the source device the data is coming from, and the destination port number indicates which application on the destination host the data is going to.

Sockets integrate IP addresses and port numbers to describe both sides of a complete communication. Sockets are expressed together with an IP address and a port number in the format: "IP address: port number", describing a particular application or service on a particular host that is involved in the communication. Sockets are used to represent a data communication in detail: an application with port number 8080 on host 192.168.0.1 sends a set of data to an application or service with port number 80 on host 192.168.0.2. This is shown in Figure 4.16.

Describing data communication with sockets is more detailed and accurate than describing communication between hosts with IP addresses alone or describing communication between different applications with port numbers alone. This is because describing the two sides of the communication

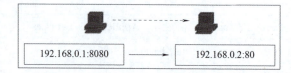

Figure 4.16 Describing the Data Communication Process Using Sockets

with IP address alone can only describe from which host the data is sent to which host, and which application on the host from which this data is transmitted cannot be expressed; if the two sides of the communication are expressed with the port number alone, it can only describe which application sends out the data to which application it is to be sent to, but it cannot describe the application on which host it is.

10. The three handshakes of the TCP—a collaborative attack puzzle

Why is the number of times the sender's and the receiver's TCP protocols communicate with each other three times when they synchronise before transmitting data? Instead of two, four, or whatever?

This comes from a classic game theory case: the coordinated attack conundrum-the general's dilemma.

The story goes: Two generals, each leading their own troops, were waiting for the enemy by ambushing them on two hills some distance apart. General A receive reliable information that the enemy had just arrived and had not yet gained a foothold. If both forces attacked the enemy together, they would be able to win; if only one force attacked, the attacker would lose.

General A had a dilemma: how to coordinate with General B and attack together?

In those days, there were no telephone or other means of communication, and the only way to pass on information was to send an informer. General A sent an informant to tell General B that the enemy was defenceless and that the two armies would "attack together at dawn". However, it may happen that the informant disappeard or was captured by the enemy. That is, although General A sent an informer to General B to convey the message of "attacking at dawn", he was not sure whether General B had received his message. In fact, even if the informant returned, General A was at a loss: how does General B know that my informant has definitely returned? If General B was not sure that the informant had returned safely, he would not attack. So General A sent the informant to General B. However, he would not be sure that this time the informant had arrived at General B's place.

No matter how many times the two generals send their informants to each other, they will not be able to coordinate. Let's optimise the problem here: assume that the environment is safe, and the two generals agree that if they receive a message from each other, they only need to confirm it to each other. Under the optimised ideal conditions (the environment is safe and the informants are not in any danger), the scenario in which the two generals send each other informants is shown in Figure 4.17.

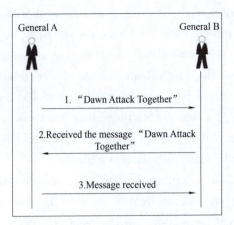

Figure 4. 17 Message Communication Process in
An Ideal State Collaborative Attack Puzzle

It can be seen that, ideally, the two generals would need to communicate at least three times to achieve synergy. The idea behind the TCP's three handshakes is the same: the sender and the receiver need to communicate at least three times to achieve synchronisation.

The scenario corresponds to the three handshakes of the TCP:

①Sender: I am ready to send (SYN);

②Receiver: Message received (ACK), I am ready to receive (SYN);

③Sender: Message received (ACK).

11. Security Issues Arising from the Three Handshakes of the TCP

If the receiver receives a SYN from the sender and returns a SYN+ACK to the sender, at which point the sender goes offline and the receiver does not receive an ACK back from the sender, the third handshake fails, as shown in Figure 4. 18.

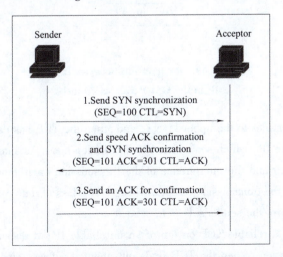

Figure 4. 18 Third Handshake Message Delivery Failure

Then, the three handshakes of the TCP are in an intermediate state, neither succeed nor fail. So, if the receiver does not receive the third handshake message from the sender within a certain period of time, the receiver's TCP protocol will resend the second handshake message (ACK+SYN) to the sender. For Linux systems, the default number of retries is 5, and the retry interval is doubled every time from 1 s. The 5 retry intervals are 1 s, 2 s, 4 s, 8 s, and 16 s, which is a total of 31 s; you have to wait for 32 s after the 5th time before you know that the 5th time has timed out as well, so it will take a total of 1 s+2 s+4 s+8 s+16 s+32 s=63 s before the receiver's TCP will disconnect the connection and release the related resources.

If the retry delay mechanism is exploited maliciously, and the sender sends a large number of TCP requests to the receiver but intentionally does not send the third handshake message (ACK), it will consume a large amount of resources on the receiver, which may ultimately result in the receiver's failing to provide normal services due to resource exhaustion.

Linux systems can use three TCP parameters to adjust the behaviour of retry delays: the_synack_retries parameter, which adjusts the number of retries; the_max_syn_backlog parameter, which adjusts the number of SYN connections; and the_abort_on_overflow parameter, which adjusts the behaviour of the connection when it exceeds the connection capacity. The behaviour when the connection capacity is exceeded.

12. The TCP is Reliable, But Then It Sends Data to the Unreliable IP for Transmission; Is the Result Reliable or Unreliable

First we look at the relationship between the TCP and IP in the OSI reference model. The TCP works at the transport layer, while the IP works at the network layer. This is shown in Figure 4.19.

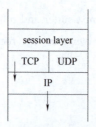

Figure 4.19 Relationship between TCP and
IP in the OSI Reference Model

The IP provides services to the upper layers, and when the TCP has data to send, it sends the data segment down to the IP, which sends it to the destination host. As analysed earlier, due to the complexity of the network and the interference of the transmission environment, it is inevitable that errors will occur during the transmission of data. If there is a loss of data, the receiver's TCP sends an ACK message to inform the sender's TCP to send it again.

This is like having a reliable TCP on top of an unreliable IP; it doesn't matter if the IP loses the data during transmission, when the TCP finds out about it, it will give the data to the IP and ask the IP to send it again until the data reaches the receiver successfully.

Project 4 Analyzing Network Communications

So even though the IP is a best-effort, unreliable protocol, it is still reliable as far as the whole process of data transfer is concerned because it is monitored by a reliable TCP.

It is similar to when we shop online. The seller sends the goods to the buyer via a courier, there is a chance that the courier company may lose the courier, at which point the seller sends another copy to ensure that the buyer gets the goods without any problem. So, it doesn't matter if the courier company is unreliable, as long as the seller is reliable, it ensures that the whole transaction is completed smoothly.

Thinking and Training

Based on the OSI network model, describe the details of the data communication process of the intranet user Laptop1 accessing the Internet network web server in Figure 4.20.

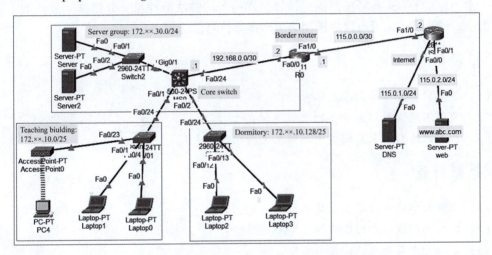

Figure 4.20 Task Topology Diagram

项目 5

排除网络故障

网络中断将会严重影响企业业务，如何快速找到并排除计算机网络故障，是本项目的重点。本项目基于 OSI 参考模型描述计算机网络各层的故障现象，介绍排除故障的工具及一般方法。学生通过练习可以掌握网络故障排除技巧。

知识目标

（1）理解网络故障的排除思路；
（2）理解不同故障排除命令的工作原理；
（3）理解网络 OSI 七层模型中不同层之间的故障表现。

技能目标

（1）掌握网络故障诊断工具及其排除方法；
（2）掌握数据链路层网络故障的诊断与排除；
（3）掌握网络层故障的诊断与排除。

相关知识

1. 网络故障排除一般思路

在网络遭遇故障时，最困难的不是修复网络故障本身，而是如何迅速地查出故障点，确定发生的原因。对于网络工程师来说，要有一个清晰的排除网络故障的思路。

根据所学过的 OSI 参考模型，从故障的实际现象出发，以网络诊断工具为手段获取诊断信息，有两种方式逐步排查故障：自下而上或者自上而下。

自下而上就是沿着 OSI 参考模型的 7 层，从物理层开始依次向上排查，逐步确定网络故障点，查找问题的根源，排除故障，恢复网络的正常运行。

自上而下则刚好相反。沿着 OSI 参考模型，从第 7 层开始，先检查应用层，层层往下，排查每一层可能的故障点，直到网络服务恢复正常。

但是不管是自下而上还是自上而下，周期都比较长。比如采用自下而上的方法，如果故障出现在应用层，因为是从物理层开始排查，则要排查完所有 7 层才可以检查到故障。自上而下的方式如果遇到物理层的故障，则有可能也会遇到这样的问题。

最常用的方法是二分法。即从中间入手，先检查网络层的故障，如果网络层没有问题，则沿着 OSI 参考模型往上检查；如果网络层有故障，则从网络层沿着 OSI 参考模型往下检查各层。

2. 常见故障点

园区网络通信故障的问题一般主要出现在物理层故障（第 1 层）、数据链路层故障（第 2 层）、网络层故障（第 3 层）等，本项目中我们采用自下而上的方法进行故障排除。

第 1 层物理层故障主要现象有硬件故障、线路故障及逻辑故障。硬件故障常见为网络设备、网卡物理本身的硬件故障，一般为设备硬件损坏、端口损坏、插头松动等；线路故障表现为网线或者光纤线路本身物理损坏；逻辑故障一般为配置错误，例如线缆两端的工作速率、工作方式不一致等问题会导致物理层故障。在 Cisco Packet Tracer 所搭建的园区网络中，物理层的故障主要表现为线缆的选择错误，线缆两端的工作速率、工作方式不一致。更换正确的线缆，把工作速率、工作方式设置为一致即可解决问题，此处暂不详细阐述。

第 2 层数据链路层故障常表现为 VLAN 协议故障。在以太网交换技术中，VLAN 技术是最为重要的技术之一，也是应用最为广泛的技术之一。VLAN 故障排除基本思路主要有：首先，分析 VLAN 是否存在且一致；其次，检查端口是否属于指定的 VLAN；再次，检查交换机两端 Trunk 端口是否配置正确；最后，检查交换机虚拟端口 SVI 是否配置正确。

常见的故障排除步骤如表 5.1 所示。

表 5.1　常见的故障排除步骤

可能的原因	判断方法	解决办法
该 VLAN 没有创建，且 VLAN 名不一致	使用 show vlan 命令检查 VLAN 是否已经存在，名字是什么	通过 VLAN ××命令进行创建；通过 name ×××修改 VLAN 名
端口没有划入指定的 VLAN 中	使用 show vlan 命令检查端口是否已经划入指定的 VLAN 中	先把端口设置为 access 模式，再把端口划入指定 VLAN 中
交换机两端 Trunk 配置有误	使用 show interface trunk 命令检查交换机两端 trunk 是否配置正确，状态是否正常	①没有配置 Trunk。把端口类型设置为 Trunk。②封装的协议有误。把交换机两端封装的协议改为一致。③允许通过的 VLAN 信息有误。允许通过对应的 VLAN
SVI 端口配置有误	使用 show ip int brief 检查交换机 SVI 端口配置是否正确	①SVI 端口配置错误。修改交换机 SVI 端口地址为 PC 的网关。②SVI 端口错误地配置到二层交换机。取消二层交换机的 SVI 配置

第 3 层网络层故障主要表现为：路由表条目不完整、NAT 配置出错等。通过使用路由追踪 Tracert 命令，检查路径走向即可解决问题。

任务 1：排除网络第 2 层故障——Vlan、Trunk、SVI （交换机虚拟端口）

任务目标：能找出网络中的第 2 层故障点并排除故障。

这里基于项目 3 的案例进行分析。为了开展本项目的学习，我们在项目的网络设备中预设了错误的配置，销售部客户端 PC 无法访问网络，需要进行故障排除。

排除网络第 2 层故障——VLAN、Trunk、SVI （交换机虚拟端口）

1. 检查客户端 PC 地址信息配置是否正确，到网关连通性是否正常

根据如图 5.1 所示的项目网络拓扑，数据的流向为：客户端 Laptop1—Switch0—MS0—R0—Internet。

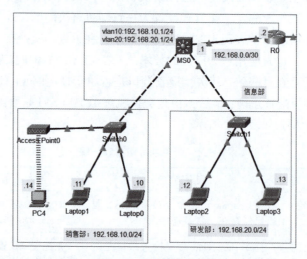

图 5.1　项目网络拓扑

客户端 PC 要访问互联网，因为属于跨网络的通信，首先需要把信息发送到网关。

首先检查计算机的地址是否配置正确。在客户端 Laptop0 中输入 ipconfig 查看 IP 地址信息，如图 5.2 所示。

```
C:\>ipconfig
FastEthernet0 Connection:(default port)

   Link-local IPv6 Address.........: FE80::201:42FF:FE15:703A
   IP Address......................: 192.168.10.10
   Subnet Mask.....................: 255.255.255.0
   Default Gateway.................: 192.168.10.1
```

图 5.2　查看客户端 IP 地址信息

查到自己的 TCP/IP 的参数：IP 地址（IP Address）是 192.168.10.10，子网掩码（Subnet Mast）是 255.255.255.0，默认网关（Default Gateway）是 192.168.10.1，这些参数都是配齐了的。一般会直接去测试客户端 PC 和网关之间的连通性，如图 5.3 所示。

```
C:\>ping 192.168.10.1

Pinging 192.168.10.1 with 32 bytes of data:

Request timed out.
Request timed out.
Request timed out.
Request timed out.

Ping statistics for 192.168.10.1:
    Packets: Sent = 4, Received = 0, Lost = 4 (100% loss),
C:\>
```

图 5.3　测试客户端 PC 和网关之间的连通性

通过如图 5.2 和图 5.3 所示的分析可见，客户端的 IP 地址信息配置正确，但是无法访问网关。

出问题要先从自己身上检查。Laptop0 和网关 ping 各自的 IP 地址，测试自身的 TCP/IP 协议栈工作是否正常，结果如图 5.4 和图 5.5 所示。

```
C:\>ping 192.168.10.10

Pinging 192.168.10.10 with 32 bytes of data:

Reply from 192.168.10.10: bytes=32 time=3ms TTL=128
Reply from 192.168.10.10: bytes=32 time<1ms TTL=128
Reply from 192.168.10.10: bytes=32 time=2ms TTL=128
Reply from 192.168.10.10: bytes=32 time=4ms TTL=128
```

图 5.4　Laptop0 主机 ping 自己的结果

```
MS0#ping 192.168.10.1

Type escape sequence to abort.
Sending 5, 100-byte ICMP Echos to 192.168.10.1, timeout is 2 seconds:
!!!!!
Success rate is 100 percent (5/5), round-trip min/avg/max = 1/5/14 ms
```

图 5.5　网关 MS0 ping VLAN 10 端口的结果

Laptop0 和网关 ping 自己的 IP 地址都没有问题，于是几乎可以判断是属于第 2 层的网络故障。而网络第 2 层的故障点经常出现在 VLAN、Trunk、SVI 的配置上，下面将逐一排除。

2. 检查接入层交换机的 VLAN 配置是否正确

首先，在交换机 Switch0 中检查 VLAN 数据库信息是否有误，用命令 show vlan brief 查询 VLAN 数据库的信息，结果如图 5.6 所示。

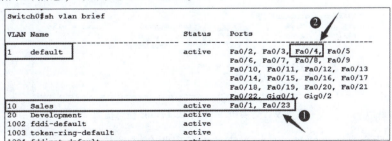

图 5.6　查看交换机 VLAN 数据库信息

这个图显示了 VLAN 10 的端口成员只有 2 个（标①处），连接计算机 Laptop0 的 Fa0/4 端口（标②处），被错误的放入了 VLAN 1 中。

在 Switch0 中修正错误信息，把 Fa0/4 端口划入 VLAN 10 中，输入命令如下：

```
Switch0(config)#int fa0/4
Switch0(config-if)#switchport mode access
Switch0(config-if)#switchport access vlan 10
Switch0(config-if)#exit
```

然后，再检查 VLAN 配置是否正确，结果如图 5.7 所示。

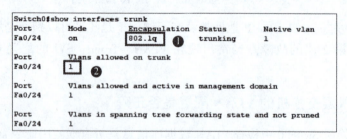

图 5.7　正确配置后的 VLAN 数据库信息

用同样的方法，检查其他接入层交换机的 VLAN 配置是否正确。

3. 检查交换机的 Trunk 端口状态是否正确

交换机 Switch0 和交换机 MS0 之间的链路为干道（Trunk），需要检查 Trunk 状态信息是否有误。在交换机特权模式输入 show interfaces trunk 命令可以看到端口 Trunk 状态信息。交换机 Switch0 和交换机 MS0 之间链路的 Trunk 状态信息如图 5.8、图 5.9 所示。

```
Switch0#show interfaces trunk
Port      Mode         Encapsulation  Status        Native vlan
Fa0/24    on           802.1q         trunking      1

Port      Vlans allowed on trunk
Fa0/24    1

Port      Vlans allowed and active in management domain
Fa0/24    1

Port      Vlans in spanning tree forwarding state and not pruned
Fa0/24    1
```

图 5.8　交换机 Switch0 端口 Trunk 状态信息

```
MS0#show int trunk
Port      Mode         Encapsulation  Status        Native vlan
Fa0/1     auto         n-802.1q       trunking      1
Fa0/2     on           802.1q         trunking      1

Port      Vlans allowed on trunk
Fa0/1     1
Fa0/2     1-1005
```

图 5.9　交换机 MS0 端口 Trunk 状态信息

通过查看两个交换机的 Trunk 状态信息，发现两个错误点：

第一个错误点：对比图 5.8 和图 5.9 标①处，可以看出，交换机 Switch0 封装的为 802.1q 协议；而交换机 MS0 封装的是 n-802.1q 协议，交换机 Switch0 和交换机 MS0 之间封装的协议不同，有误。

第二个错误点：如图 5.8 和图 5.9 所示的标②处表明，交换机 Switch0 和交换机 MS0 之间的 Trunk 链路仅允许通过的 VLAN 为 1，有误。

找到错误点后，需要进行修正。对于第一个错误点，在交换机 MS0 中，把 Trunk 干道封装的协议修改为 802.1q，保持两端交换机一致，命令如下：

> MS0(config)#interface fa0/1
> MS0(config-if)#switchport trunk encapsulation dot1q   //三层交换机 MS0 Trunk 封装协议改为 802.1q

对于第二个错误点，需要在交换机 Switch0 和交换机 MS0 中添加允许通过的 VLAN 10，命令如下：

> Switch0(config)#int fa0/24
> Switch0(config-if)#switchport trunk allowed vlan 10   //添加允许 VLAN 10 通过 Trunk 干道

完成后，再次检查 Trunk 配置信息，结果如图 5.10 所示。

图 5.10　修正后的交换机 MS0 端口 Trunk 状态信息

4. 检查交换机的 SVI 配置是否正确

交换机的 SVI（交换机虚拟端口）一般配置在三层交换机中，充当 PC 网关的角色。PC 机访问外网时，首先把数据包送到三层交换机的 SVI 端口，然后再由三层交换机转发出去。如果 SVI 端口的地址不是 PC 网关的地址，或者 SVI 端口错误地配置到了二层交换机上，就会导致网络不通。

检查交换机的 SVI 端口配置正确与否的命令如下：

> MS0#show ip interface brief   //查看三层交换机的 SVI 端口配置

通过检查二层交换机和三层交换机的 SVI 端口配置，结果如图 5.11 和图 5.12 所示。

图 5.11　二层交换机 Switch0 的 SVI 端口信息

```
MS0#show ip interface brief
Interface              IP-Address      OK? Method Status                Protocol
            此处省略部分非关键性信息
Vlan1                  unassigned      YES unset  administratively down down
Vlan10                 192.168.10.1    YES manual administratively down down    ❶
Vlan20                 192.168.20.1    YES manual up                    up
```

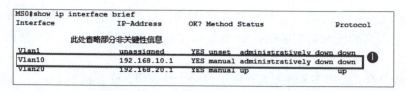

图 5.12　三层交换机 MS0 的 SVI 端口信息

通过图 5.11 和图 5.12 可以发现，标①处本来应该配置在三层交换机 MS0 的 SVI 端口信息被错误地配置到了二层交换机 Switch0 上，而二层交换机是无法实现网关功能的，且配置了与网关一样的 IP 地址，会和三层交换机的真正的网关的 IP 地址冲突；三层交换机 MS0 的 SVI 端口被人为地关闭了。

修正方法：把二层交换机 Switch0 的 SVI 端口地址删除，把三层交换机 MS0 的 SVI 端口启用即可，命令如下：

```
Switch0(config)#interface vlan 10        //进入二层交换机 Switch0 的 VLAN 10 端口
Switch0(config-if)#no ip address         //删除原有的 IP 地址配置
```

```
MS0(config)#interface vlan 10        //进入三层交换机 MS0 的 VLAN 10 端口
MS0(config-if)#no shut        //启用该 SVI 端口
```

完成后，再次检查交换机的 SVI 配置信息，检查结果如图 5.13、图 5.14 所示。

```
MS0#show ip interface brief
Interface              IP-Address      OK? Method Status                Protocol
            此处省略部分非关键性信息
Vlan1                  unassigned      YES unset  administratively down down
Vlan10                 192.168.10.1    YES manual up                    up
Vlan20                 192.168.20.1    YES manual up                    up
```

图 5.13　三层交换机 MS0 的 SVI 端口信息

```
Switch0#show ip interface brief
Interface              IP-Address      OK? Method Status                Protocol
            此处省略部分非关键性信息
Vlan1                  unassigned      YES manual administratively down down
Vlan10                 unassigned      YES manual up                    up
```

图 5.14　二层交换机 Switch0 的 SVI 端口信息

同样的方法，排除了交换机 MS0 和 Switch1 的二层网络常见的故障点之后，再测试下各个 PC 的网络连通性，此时网络已经正常，PC 能正常访问互联网，如图 5.15 所示。

图 5.15　PC 能正常访问互联网

课堂练习：完成电子资源库中配套的二层故障网络模型，实现内网用户能访问 Internet 资源。

素质拓展：掌握网络故障排除技巧是一名网络工程师的必备技能。但由于网络故障有很多不确定性，故障排除容易让初学者感到困惑，必须勇敢去面对。纵观人类社会发展史，任何一种理

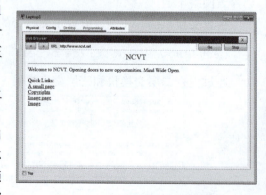

想的实现都不是轻而易举的，必然会遇到各种各样的困难和波折，充满着艰险和坎坷。大学生要正确认识、处理生活中各种各样的困难和问题，保持认真务实、乐观向上、积极进取的人生态度。

任务 2：排除网络第 3 层故障——路由追踪 Tracert、NAT

任务目标：能找出网络中的 3 层故障点并排除故障。

这里以项目 3 的案例进行分析，任务网络拓扑如图 5.16 所示。由于网络设备被进行了错误的配置，客户端 PC 无法使用网络，需要进行故障排除。

排除网络第 3 层故障——路由追踪 TRACERT、NAT

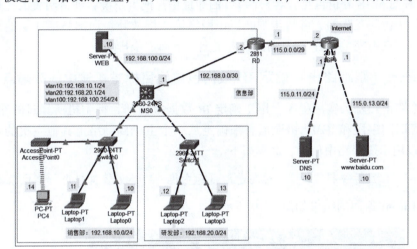

图 5.16 任务网络拓扑

1. 检查客户端 PC 地址信息配置是否正确，到网关连通性是否正常

首先，通过 ipconfig 命令查询客户机的 IP 地址信息，然后用 ping 命令检测 PC 到网关的连通性是否正常，查看客户端 IP 地址信息结果如图 5.17 所示，查看客户端 PC 与网关连通性结果如图 5.18 所示。

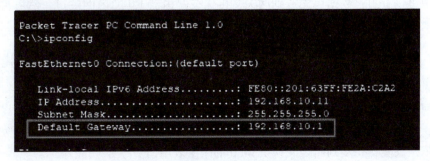

图 5.17 查看客户端 IP 地址信息结果

```
C:\>ping 192.168.10.1

Pinging 192.168.10.1 with 32 bytes of data:

Reply from 192.168.10.1: bytes=32 time<1ms TTL=255
Reply from 192.168.10.1: bytes=32 time<1ms TTL=255
Reply from 192.168.10.1: bytes=32 time=1ms TTL=255
Reply from 192.168.10.1: bytes=32 time<1ms TTL=255

Ping statistics for 192.168.10.1:
    Packets: Sent = 4, Received = 4, Lost = 0 (0% loss),
Approximate round trip times in milli-seconds:
    Minimum = 0ms, Maximum = 1ms, Average = 0ms
```

图 5.18 查看客户端 PC 与网关连通性结果

通过图 5.17 和图 5.18 分析可见，客户端 PC 的地址信息配置正确，到达网关的连通性正常，但是仍然无法访问互联网，可初步判断属于网络第 3 层故障。而网络第 3 层的故障点经常出现在路由信息不完整或者 NAT 的配置有误上，下面逐步排除。

2. 使用 Tracert（跟踪路由）确定 IP 数据包访问目标主机途中的故障点

Tracert 是一个路由跟踪的实用程序，用于确定 IP 数据报访问目标主机所经过的路径，Tracert 命令用生存时间字段和 ICMP 错误消息来确定从一个主机到网络上其他主机的路由 IP 地址，以此可以用于追踪路由信息，命令如下：

```
C:\>tracert 目标主机 IP    //对发送到目标主机的 ICMP 数据包进行路由跟踪
```

网络正常时 Tracert 返回的结果如图 5.19 所示。

```
C:\>tracert 115.0.13.10

Tracing route to 115.0.13.10 over a maximum of 30 hops:

  1    0 ms    0 ms    0 ms    192.168.10.1
  2    0 ms    0 ms    0 ms    192.168.0.2
  3    2 ms    1 ms    1 ms    115.0.0.2
  4    0 ms    1 ms    0 ms    115.0.13.10

Trace complete.
```

图 5.19 网络正常时 Tracert 返回的结果

从图中可以看出，数据包先送到网关，再送往路由器 R0 的内网端口，然后再送到 ISP 的网关，最后到达目标主机 115.0.13.10。每到一个网络节点，都会有一个应答，并记录下来。

然而，此处网络通信并不正常，在客户端 PC 上 Tracert 目标主机，返回如图 5.20 所示的异常结果。

该如何进行诊断呢？

首先，分析输出信息，可以看出数据在到达网关后就不可达了，然后重新发给网关，再超时，依次反复，初步判断是交换机 MS0 的路由表出错。

其次，在交换机上查看路由表，命令如下：

```
MS0#show ip route    //查看三层交换机的路由表
```

图 5.20 网络异常时 Tracert 返回的结果

从图 5.21 所示的交换机 MS0 的路由表中看出，交换机 MS0 的路由表仅有直连路由，并没有到达目标网络的路由。

图 5.21 交换机 MS0 的路由表

再次，检查交换机的 MS0 的配置文件，检查三层交换机 MS0 的路由功能是否开启，路由协议是否正确配置，检查的命令如下：

MS0#show run   //查看三层交换机的配置

交换机 MS0 的配置文件信息如图 5.22 所示，①处表明了交换机开启了路由功能，是正确的；②处表示交换机配置了一条默认路由，下一跳指向了 192.168.0.254 这个 IP 地址，对照网络拓扑图，发现此处的下一跳 IP 地址写错了，是故障点。

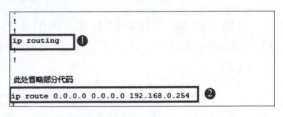

图 5.22 交换机 MS0 的配置文件信息

最后，删除错误的路由信息，添加新的路由条目，命令如下：

MS0(config)#ip route 0.0.0.0 0.0.0.0 192.168.0.254   //删除错误的默认路由
MS0(config)#ip route 0.0.0.0 0.0.0.0 192.168.0.2   //设置默认路由下一跳为 192.168.0.2

修改了错误的默认路由后,重新查看交换机的路由表,就能显示出正确的下一跳地址了,交换机 MS0 正确的路由表如图 5.23 所示。

```
MS0#show ip route
Codes: C - connected, S - static, I - IGRP, R - RIP, M - mobile, B - BGP
       D - EIGRP, EX - EIGRP external, O - OSPF, IA - OSPF inter area
       N1 - OSPF NSSA external type 1, N2 - OSPF NSSA external type 2
       E1 - OSPF external type 1, E2 - OSPF external type 2, E - EGP
       i - IS-IS, L1 - IS-IS level-1, L2 - IS-IS level-2, ia - IS-IS inter area
       * - candidate default, U - per-user static route, o - ODR
       P - periodic downloaded static route

Gateway of last resort is 192.168.0.2 to network 0.0.0.0

     192.168.0.0/30 is subnetted, 1 subnets
C       192.168.0.0 is directly connected, FastEthernet0/24
C    192.168.10.0/24 is directly connected, Vlan10
C    192.168.20.0/24 is directly connected, Vlan20
S*   0.0.0.0/0 [1/0] via 192.168.0.2
```

图 5.23　交换机 MS0 正确的路由表

3. 解决路由器 NAT 故障

修正了三层交换机 MS0 的路由条目后,客户机仍然无法访问互联网。

首先,按第 2 步的方法通过 Tracert 路由跟踪发现,数据包已经能正确到达路由器 R0 的端口,但是无法到达下一跳,如图 5.24 所示。

```
C:\>tracert 115.0.13.10

Tracing route to 115.0.13.10 over a maximum of 30 hops:

  1    1 ms     0 ms     0 ms   192.168.10.1
  2    1 ms     1 ms     0 ms   192.168.0.2
  3    *        *        *      Request timed out.
  4    *        *        *      Request timed out.
  5    *        *        *      Request timed out.
  6    *        *        *      Request timed out.
  7    *        *        *      Request timed out.
  8    *        *        *      Request timed out.
  9    *        *        *      Request timed out.
```

图 5.24　重新 Tracert 路由跟踪

在路由器 R0 上检查路由表,如图 5.25 所示,①处为路由器返回内网 VLAN 10、VLAN 20 的静态路由,②处为路由器到达 ISP 网关的默认路由,路由信息及下一跳地址均正确无误。

```
R0#show ip route
Codes: C - connected, S - static, I - IGRP, R - RIP, M - mobile, B - BGP
       D - EIGRP, EX - EIGRP external, O - OSPF, IA - OSPF inter area
       N1 - OSPF NSSA external type 1, N2 - OSPF NSSA external type 2
       E1 - OSPF external type 1, E2 - OSPF external type 2, E - EGP
       i - IS-IS, L1 - IS-IS level-1, L2 - IS-IS level-2, ia - IS-IS inter area
       * - candidate default, U - per-user static route, o - ODR
       P - periodic downloaded static route

Gateway of last resort is 115.0.0.2 to network 0.0.0.0

     115.0.0.0/29 is subnetted, 1 subnets
C       115.0.0.0 is directly connected, FastEthernet1/0
     192.168.0.0/30 is subnetted, 1 subnets
C       192.168.0.0 is directly connected, FastEthernet0/0
S    192.168.10.0/24 [1/0] via 192.168.0.1         ❶
S    192.168.20.0/24 [1/0] via 192.168.0.1
S*   0.0.0.0/0 [1/0] via 115.0.0.2                 ❷
```

图 5.25　路由器 R0 的路由表

同时，在路由器 R0 上 ping 运营商的网关地址是正常的，如图 5.26 所示。

```
R0#ping 115.0.0.2
Type escape sequence to abort.
Sending 5, 100-byte ICMP Echos to 115.0.0.2, timeout is 2 seconds:
!!!!!
Success rate is 100 percent (5/5), round-trip min/avg/max = 0/0/1 ms
```

图 5.26　检测路由器 R0 到 ISP 网关的连通性

其次，利用项目 3 所学的知识进行分析可知，实现内网用户访问互联网服务器的方法是使用端口地址转换（PAT）技术。排除了路由器的路由配置及连通性问题，就可判断故障点在 NAT 了。

如图 5.27 所示，在路由器 R0 上查看配置信息，可发现有 4 处错误点：

①处为端口 FastEthernet0/0 为内网端口，错误地配置成外网端口。
②处为端口 FastEthernet1/0 本该为外网端口，错误地配置成内网端口。
③处为与后面的 ACL 定义的编号，错误地写成了 2。
④处为端口 FastEthernet0/0 不是连接外网的端口。
这 4 个故障点均是初学者较容易犯的错误。

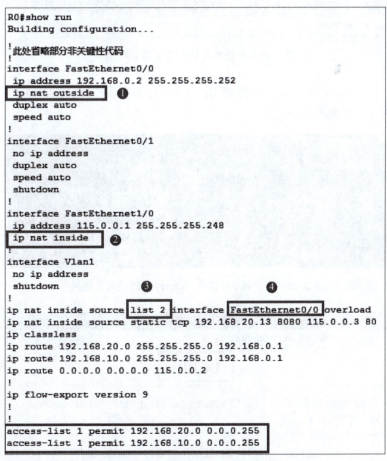

图 5.27　路由器 R0 的配置故障点

再次，对照错误点，把原有错误的配置删除，重新修正路由器 R0 的配置，正确配置后的代码如图 5.28 所示。

```
R0#show run
Building configuration...
!
此处省略部分非关键性代码
!
interface FastEthernet0/0
 ip address 192.168.0.2 255.255.255.252
 ip nat inside
 duplex auto
 speed auto
!
interface FastEthernet0/1
 no ip address
 duplex auto
 speed auto
 shutdown
!
interface FastEthernet1/0
 ip address 115.0.0.1 255.255.255.248
 ip nat outside
!
interface Vlan1
 no ip address
 shutdown
!
ip nat inside source list 1 interface FastEthernet1/0 overload
ip nat inside source static tcp 192.168.20.13 8080 115.0.0.3 80
ip classless
ip route 192.168.20.0 255.255.255.0 192.168.0.1
```

图 5.28 修正故障后的路由器 R0 配置

最后，在客户端 PC 通过 Tracert 跟踪访问互联网，使用浏览器访问 baidu 服务器，均已成功，如图 5.29、图 5.30 所示。

```
C:\>tracert 115.0.13.10

Tracing route to 115.0.13.10 over a maximum of 30 hops:

  1   0 ms      1 ms      0 ms      192.168.10.1
  2   0 ms      1 ms      0 ms      192.168.0.2
  3   2 ms      0 ms      0 ms      115.0.0.2
  4   0 ms      0 ms      0 ms      115.0.13.10

Trace complete.
```

图 5.29 客户端 PC 使用 Tracert 跟踪访问互联网

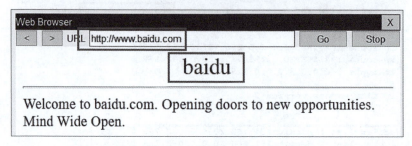

图 5.30 客户端 PC 使用浏览器访问 baidu 服务

课堂练习：完成电子资源库中配套的 3 层故障网络模型，实现内网用户能访问 Internet 资源。

素质拓展：3 层网络故障通过 Tracert、ping 等命令，结合各种工具进行故障定位和排查，可快速积累故障排除经验，提升技术水平。我们要善于利用各种工具提高解决问题的效率。马克思主义揭示了事物的本质、内在联系及发展规律，是"伟大的认识工具"，是人们观察世界、分析问题的有力思想武器。大学生只有确立马克思主义的科学信仰，才能真正确立崇高的理想信念，在错综复杂的社会现象中看清本质、明确方向，为服务人民、奉献社会作出更大的贡献。

小结与拓展

1. 网络故障排除经验总结

网络故障一般分为两大类：连通性问题和性能问题。
（1）连通性问题。
主要出现在硬件、系统、电源、传输介质故障，IP 地址配置错误等。
（2）性能问题。
主要出现在网络拥塞、到目的地不是最佳路径、转发异常、路由环路等。
故障排除需要一步一步找出故障原因并解决，它的基本思想是系统地将由故障可能的原因所构成的一个大集合缩减（或隔离）成几个小的子集，从而使问题的复杂度迅速下降。故障排除时有序的思路有助于解决所遇到的任何困难，如图 5.31 所示是一般网络故障解决的处理流程。

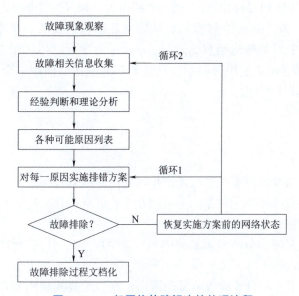

图 5.31　一般网络故障解决的处理流程

2. 日常使用的网络故障排查命令总结

（1）ping。

在网络故障排除中，使用最频繁的是 ping 命令，它不仅可以检查网络是否连通，还有益于分析判断网络故障。其常用方法有：

①ping 本机 IP。

本机始终应该对 ping 命令作出应答，如没有，则表示本地 TCP/IP 协议栈存在问题。

②ping 网关 IP。

可以检查本机与本地网络连接是否正常。

③ping 远程 IP。

如果收到 4 个应答，表示网络正常。

④ping 局域网内其他 IP。

数据包会经过网卡及网缆到达其他机器后再返回。收到回送应答表明本地网络中运行正确；但若收到 0 个回送应答，那么表示子网掩码不正确或网卡配置错误或网线有问题。

⑤ping 127.0.0.1。

127.0.0.1 环回地址是为了检查本地的 TCP/IP 协议是否设置正确。

⑥ping www.xxx.com（如 www.baidu.com）

检查 DNS 能否解析出正确的 IP 地址，及到达目标服务器是否正常。

（2）tracert。

tracert 是为了探测源节点到目的节点之间数据报文经过的路径。如果有网络连通性问题，可以使用 tracert 命令来检查到达的目标 IP 地址的路径并记录结果，tracert 命令显示用于将数据包从计算机传递到目标位置的一组 IP 路由器，以及每个跃点所需的时间。如果数据包不能传递到目标，tracert 命令将显示成功转发数据包的最后一个路由器，当数据报从我们的计算机经过多个网关传送到目的地时，tracert 命令可以用来跟踪数据传输过程中使用的路由（路径）。tracert 的使用很简单，只需要在 tracert 后面跟一个 IP 地址或域名即可判断网络在哪个环节出了问题。

思考与训练

1. 简答题

（1）局域网常见的故障有哪些？故障产生可能的原因都有哪些？

（2）常用的网络故障排除的网络命令有哪几个？各自有什么功能？

（3）局域网故障诊断的方法主要有哪些？

（4）网络出现故障时，分析步骤如何进行？请用自己的语言进行描述。

2. 实验题

完成配套电子资源库项目 5 的故障排除网络模型，最终实现内网用户能访问到 Internet 资源，同时允许外部网络用户能访问到内网服务器。

Project 5

Troubleshooting the network

Network disruption will seriously affect the enterprise's business How to quickly find and troubleshoot computer network faults is the focus of this project. This Project based on the OSI reference model, this chapter describes the failure phenomena of each layer of the computer network, introduces troubleshooting tools and general methods. Through practice, you can master network troubleshooting skills.

Learning Objectives

(1) Understand network troubleshooting ideas;
(2) Understand how different troubleshooting commands work;
(3) Understand the behavior of failures between different layers in the seven-layer OSI model of networking.

Skill Objectives

(1) Knowledge of network troubleshooting tools and how to troubleshoot them;
(2) Master the diagnosis and troubleshooting of data link layer network faults;
(3) Master the diagnosis and troubleshooting of network layer faults.

Relevant Knowledge

1. General Ideas for Network Troubleshooting

When a network encounters a failure, the most difficult part is not fixing the network failure itself, but how to quickly identify the point of failure and determine why it happens. It is important for network engineers to have a clear idea of how to troubleshoot a network.

Based on the OSI reference model that you have learned, there are two ways to step-by-step troubleshooting from the actual phenomenon of the fault, using network diagnostic tools as a means to obtain diagnostic information: bottom-up or top-down.

The bottom-up way is along the seven layers of the OSI reference model, starting from the physical layer upwards in order to gradually determine the network failure point, find the root cause of the problem, troubleshoot, and restore the normal operation of the network.

The top-down way is just the opposite. Along the OSI network reference model, starting at layer 7, the application layer is examined first, and the works down the layers, each layer is examined for possible points of failure until the network service is restored to normal.

But whether it is bottom-up or top-down, the cycle time is longer. For example, with the bottom-up approach, if the fault occurs in the application layer, because the troubleshooting starts from the physical layer, it will take all 7 layers to check the fault. The top-down approach is likely to encounter the same problem if it encounters a fault in the physical layer.

The most commonly-used method is the dichotomous method. That is, starting in the middle, the network layer is checked for faults first, and if the network layer is fine, the layers are checked up along the OSI reference model; if the network layer is faulty, the layers are checked down along the OSI reference model from the network layer.

2. Common Fault Points

Problems of campus network communication failures generally occur mainly in physical layer failures (Layer 1), data link layer failures (Layer 2), and network layer failures (Layer 3), etc. In this project, we use a bottom-up approach to troubleshooting.

The first layer of physical layer failures are hardware failures, line failures and logical failures. Hardware failures are common for network equipments, network card physical hardware failures. The failures are usually equipment hardware damage, port damage, loose plugs, etc. ; line failures are the physical damage manifested in the network cable or fiber optic line; logical failures are generally the wrong configurations, such as the two ends of the cable working rate, the working mode inconsistent with the problem will lead to physical layer failures. In the park network built by Cisco Packet Tracer, the physical layer failure is mainly manifested as the cable selection errors, cable ends of the working rate, working mode is inconsistent, replace the correct cable, the working rate, working mode set to the same can solve the problem, this will not be elaborated in detail.

Layer 2 data link layer failures are often manifested as VLAN protocol failures. In Ethernet switching technology, the VLAN technology is one of the most important technologies and one of the most widely-used technologies. The basic ideas of VLAN troubleshooting are: first, analyze whether the VLAN exists and is consistent; second, check whether the port belongs to the specified VLAN; then, check whether the Trunk ports at both ends of the switch are configured correctly; finally, check whether the virtual ports of the switch are configured correctly; and lastly, check whether the SVIs of the switch are configured correctly. Finally, check whether the virtual ports of the switch are configured correctly.

Common troubleshooting steps are shown in Table 5.1.

Project 5 Troubleshooting the network

Table 5.1 VLAN Troubleshooting Step-by-Step Chart

Possible Causes	Method of Judgment	Method Settle An Issue
This VLAN was not created and the VLAN names are inconsistent	Use the show vlan command to check if the VLAN already exists and what the name is	Create via the VLAN XX command; modify the VLAN name via name XXX
The port is not classified in the specified VLAN	Use the show vlan command to check whether the port has been assigned to the specified VLAN	Set the port to access mode first, and then assign the port to a specific VLAN
Trunk is incorrectly configured at both ends of the switch	Use the show interface trunk command to check whether trunk is configured correctly at both ends of the switch and whether the status is normal	①Trunk is not configured. Set the port type to Trunk. ②Incorrect protocol for encapsulation. Change the protocol encapsulated on both ends of the switch to be consistent. ③Incorrect information about the VLAN allowed to pass through. Allowed to pass through the corresponding VLAN
SVI ports are incorrectly configured	Use show ip int brief to check that the switch SVI ports are configured correctly	①SVI port configuration error. Modify the switch SVI port address to the PC's gateway. ②SVI ports misconfigured to Layer 2 switches. Cancle SVI configuration for Layer 2 switches

Layer 3 network layer failures are mainly characterized by incomplete routing table entries, NAT configuration errors, and so on. The problem can be solved by using the Tracert command to check the path direction.

Task 1: Troubleshooting Network Layer 2—Vlan, Trunk, SVI (Switch Virtual Ports)

Task Goal: Be able to identify and troubleshoot Layer 2 points of failure in a network

The case study based on project 3 is analyzed here. In order to carry out this project, we have preconfigured the network devices in the project with a wrong configuration, and the client PCs in the sales department are unable to access the network and need to be troubleshooted.

Troubleshooting Network Layer 2— Vlan,Trunk,SVI (Switch Virtual Ports)

1. Check Whether the Client PC Address Information Is Configured Correctly and Whether the Connectivity to the Gateway Is Normal

According to the network topology in Figure 5.1, the flow of data is: client Laptop1—Switch0—MS0—R0—Internet.

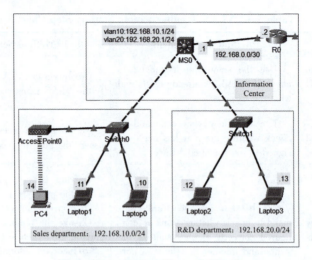

Figure 5.1 Project Network Topology

For a client PC to access the Internet, as it is a communication across networks, it first needs to send the information to the gateway.

First check that whether the computer's address is configured correctly. Type ipconfig in client Laptop0 to view the IP address information, as shown in Figure 5.2.

```
C:\>ipconfig

FastEthernet0 Connection:(default port)

   Link-local IPv6 Address.........: FE80::201:42FF:FE15:703A
   IP Address......................: 192.168.10.10
   Subnet Mask.....................: 255.255.255.0
   Default Gateway.................: 192.168.10.1
```

Figure 5.2 View Client IP Address Information

Check your TCP/IP parameters: IP Address is 192.168.10.10, Subnet Mask (Subnet Mast) is 255.255.255.0, Default Gateway (Default Gateway) is 192.168.10.1, and all of these parameters are matched. You would normally go straight to testing the connectivity between the client and the gateway. As shown in Figure 5.3.

```
C:\>ping 192.168.10.1

Pinging 192.168.10.1 with 32 bytes of data:

Request timed out.
Request timed out.
Request timed out.
Request timed out.

Ping statistics for 192.168.10.1:
    Packets: Sent = 4, Received = 0, Lost = 4 (100% loss),

C:\>
```

Figure 5.3 Test Client PC and Gateway Connectivity

Analysis of Figure 5.2 and Figure 5.3 shows that the client's IP address information is configured correctly, but it cannot access the gateway.

Project 5 Troubleshooting the network

Problems have to be checked from themselves first. laptop0 and the gateway ping their respective IP addresses to test whether their own TCP/IP stacks are working properly, and the results are shown in Figures 5.4 and 5.5.

```
C:\>ping 192.168.10.10

Pinging 192.168.10.10 with 32 bytes of data:

Reply from 192.168.10.10: bytes=32 time=3ms TTL=128
Reply from 192.168.10.10: bytes=32 time<1ms TTL=128
Reply from 192.168.10.10: bytes=32 time=2ms TTL=128
Reply from 192.168.10.10: bytes=32 time=4ms TTL=128
```

Figure 5.4 Results of Laptop0 Host Pinging Itself

```
MS0#ping 192.168.10.1

Type escape sequence to abort.
Sending 5, 100-byte ICMP Echos to 192.168.10.1, timeout is 2 seconds:
!!!!!
Success rate is 100 percent (5/5), round-trip min/avg/max = 1/5/14 ms
```

Figure 5.5 Results of Gateway MS0 Pinging VLAN 10 Port

Laptop0 and the gateway pinging their IP addresses are not a problem, so it can almost be judged to be a Layer 2 network failure. The second layer of the network failure points often appear in the VLAN, Trunk, SVI configuration, the following will be exclude one by one.

2. Check That the VLAN Configuration of The Access Layer Switch Is Correct

First, check whether the VLAN database information is incorrect in Switch0. Use the command show VLAN brief to query the VLAN database information, and the result is shown in Figure 5.6.

```
Switch0#sh vlan brief

VLAN Name                             Status    Ports
---- -------------------------------- --------- -------------------------------
1    default                          active    Fa0/2, Fa0/3, Fa0/4, Fa0/5
                                                Fa0/6, Fa0/7, Fa0/8, Fa0/9
                                                Fa0/10, Fa0/11, Fa0/12, Fa0/13
                                                Fa0/14, Fa0/15, Fa0/16, Fa0/17
                                                Fa0/18, Fa0/19, Fa0/20, Fa0/21
                                                Fa0/22, Gig0/1, Gig0/2
10   Sales                            active    Fa0/1, Fa0/23
20   Development                      active
1002 fddi-default                     active
1003 token-ring-default               active
```

Figure 5.6 Viewing the Switch VLAN Database Information

This figure shows that there are only two port members of VLAN 10 (labeled ①), and the Fa0/4 port (labeled ②) which connects to computer Laptop0 has been incorrectly placed in VLAN 1.

Correct the error message in Switch0 to assign port Fa0/4 to VLAN 10. Enter the following command:

```
Switch0(config)#int fa0/4
Switch0(config-if)#switchport mode access
Switch0(config-if)#switchport access vlan 10
Switch0(config-if)#exit
```

Then, the VLAN configuration is checked for correctness, and the result is shown in Figure 5.7.

```
Switch0#show vlan brief
VLAN Name                             Status    Ports
---- -------------------------------- --------- -------------------------------
1    default                          active    Fa0/2, Fa0/3, Fa0/5, Fa0/6
                                                Fa0/7, Fa0/8, Fa0/9, Fa0/10
                                                Fa0/11, Fa0/12, Fa0/13, Fa0/14
                                                Fa0/15, Fa0/16, Fa0/17, Fa0/18
                                                Fa0/19, Fa0/20, Fa0/21, Fa0/22
                                                Gig0/1, Gig0/2
10   Sales                            active    Fa0/1, Fa0/4, Fa0/23
1002 fddi-default                     active
1003 token-ring-default               active
1004 fddinet-default                  active
1005 trnet-default                    active
```

Figure 5.7　VLAN Database Information After Correct Configuration

Using the same method, check whether the VLAN configuration of the other access layer switches is correct.

3. Check that the switch's Trunk port status is correct

The link between switch Switch0 and switch MS0 is trunked (Trunk) and you need to check the Trunk status information for errors. Enter the show interfaces trunk command in switch privileged mode to see the port Trunk status information. The Trunk status information for the link between Switch0 and Switch MS0 is shown in Figure 5.8 and Figure 5.9.

```
Switch0#show interfaces trunk
Port      Mode         Encapsulation  Status        Native vlan
Fa0/24    on           802.1q    ①    trunking      1

Port      Vlans allowed on trunk
Fa0/24    1    ②

Port      Vlans allowed and active in management domain
Fa0/24    1

Port      Vlans in spanning tree forwarding state and not pruned
Fa0/24    1
```

Figure 5.8　Switch Switch0 Port Trunk Status Information

```
MS0#show int trunk
Port      Mode         Encapsulation  Status        Native vlan
Fa0/1     auto         n-802.1q       trunking      1
Fa0/2     on           802.1q    ①    trunking      1

Port      Vlans allowed on trunk
Fa0/1     1    ②
Fa0/2     1-1005
```

Figure 5.9　Switch MS0 Port Trunk Status Information

By looking at the Trunk status information for both switches, two points of error are found:

The first point of error: comparing Figure 5.8 and Figure 5.9 at label ①, it can be seen that Switch0 encapsulates the 802.1q protocol; while Switch MS0 encapsulates the n-802.1q protocol, and the encapsulated protocols are different between Switch0 and Switch MS0, which is in error.

The second point of error: Figure 5.8 and Figure 5.9, labeled ②, indicate that the Trunk link between Switch0 and Switch MS0 is only allowed to pass through with a VLAN of 1. There is an error.

Once the error points are found, they need to be corrected. For the first point of error, in switch MS0, change the protocol encapsulated by the Trunk trunk to 802.1q to keep the same switch at both ends, with the following command:

```
MS0(config)#interface fa0/1
MS0(config-if)#switchport trunk encapsulation dot1q //Layer 3 switch MS0 Trunk encapsulation protocol changed to 802.1q
```

For the second point of error, you need to add VLAN 10 to switch Switch0 and switch MS0 that is allowed to pass through with the following command:

```
Switch0(config)#int fa0/24
Switch0(config-if)#switchport trunk allowed vlan 10 //add allow VLAN10 to pass through Trunk trunks
```

When finished, check the Trunk configuration information again, and the result is shown in Figure 5.10:

```
MS0#show int trunk
Port       Mode          Encapsulation  Status      Native vlan
Fa0/1      on            802.1q         trunking    1
Fa0/2      on            802.1q         trunking    1

Port       Vlans allowed on trunk
Fa0/1      10
Fa0/2      1-1005
```

Figure 5.10 Corrected Switch MS0 Port Trunk Status Information

4. Check That the Switch's SVI Configuration Is Correct

The SVI (Switch Virtual Interface) of a switch is usually configured in a Layer 3 switch to act as a PC gateway. When a PC accesses the extranet, it first sends the packets to the SVI port of the Layer 3 switch, and then the Layer 3 switch forwards them out. If the address of the SVI port is not the address of the PC gateway, or the SVI port is misconfigured to the Layer 2 switch, the network will not be available.

The commands to check whether the SVI ports of the switch are configured correctly are as follows:

```
MS0#show ip interface brief //view the SVI port configuration of the Layer 3 switch
```

By checking the SVI port configurations of the Layer 2 switch and Layer 3 switch, the results are shown in Figures 5.11 and 5.12.

Figure 5.11 and Figure 5.12 show that the SVI port information that should have been configured on Layer 3 switch MS0 is wrongly configured on Layer 2 switch Switch0, which is unable

```
Switch0#show ip interface brief
Interface              IP-Address      OK? Method Status              Protocol
    Some non-critical information is omitted here

GigabitEthernet0/1     unassigned      YES manual down                down
GigabitEthernet0/2     unassigned      YES manual down                down
Vlan1                  unassigned      YES manual administratively down down
Vlan10                 192.168.10.1    YES manual up                  up
```

Figure 5.11 SVI Port Information For Layer 2 Switch Switch0

```
MS0#show ip interface brief
Interface              IP-Address      OK? Method Status              Protocol
    Some non-critical information is omitted here

Vlan1                  unassigned      YES unset  administratively down down
Vlan10                 192.168.10.1    YES manual administratively down down
Vlan20                 192.168.20.1    YES manual up                  up
```

Figure 5.12 SVI Port Information For Layer 3 Switch MS0

to realize the function of a gateway and is configured with the same IP address as the gateway, which will conflict with the IP address of the real gateway of the Layer 3 switch; and the SVI port of Layer 3 switch MS0 is artificially closed. SVI port of the Layer 3 switch MS0 is artificially closed.

Correction method: remove the SVI port address of Layer 2 switch Switch0 and enable the SVI port of Layer 3 switch MS0, the command is as follows:

```
Switch0(config)#interface vlan 10  //Enter VLAN10 port of Layer 2 switch Switch0
Switch0(config-if)# no ip address  //delete the original IP address configuration
```

```
MS0(config)#interface vlan 10  //Enter VLAN10 port of Layer 3 switch MS0
MS0(config-if)#no shut  //enable this SVI port
```

After completion, check the SVI configuration information of the switch again, and the results are shown in Figure 5.13 and Figure 5.14.

```
MS0#show ip interface brief
Interface              IP-Address      OK? Method Status              Protocol
    Some non-critical information is omitted here

Vlan1                  unassigned      YES unset  administratively down down
Vlan10                 192.168.10.1    YES manual up                  up
Vlan20                 192.168.20.1    YES manual up                  up
```

Figure 5.13 SVI Port Information For Layer 3 Switch MS0

```
Switch0#show ip interface brief
Interface              IP-Address      OK? Method Status              Protocol
    Some non-critical information is omitted here

Vlan1                  unassigned      YES manual administratively down down
Vlan10                 unassigned      YES manual up                  up
```

Figure 5.14 SVI Port Information For Layer 2 Switch Switch0

In the same way, after eliminating the common points of failures in the Layer 2 network of switch MS0 and Switch1, then test the network connectivity of each PC, at this time, the network has been normal, the PC successfully accessing the Internet results are shown in Figure 5.15.

Exercise: complete the Layer 2 Fault Network Model that accompanies the eResource Library to enable that intranet users to access Internet resources.

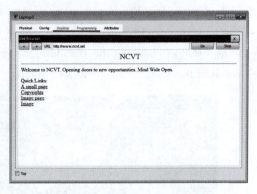

Figure 5.15 PC can access the Internet normally

Competence Expansion: Mastering network troubleshooting skills is an essential skill for a network engineer. However, due to many uncertainties of network troubleshooting, troubleshooting tends to make beginners feel confused and must be faced bravely. Throughout the history of the development of human society, the realization of any kind of ideal is not easy, and will inevitably encounter a variety of difficulties and twists and turns, full of difficulties and bumps. College students should correctly recognize and deal with all kinds of difficulties and problems in life, and maintain a conscientious and pragmatic, optimistic, positive and enterprising attitude towards life.

Task 2: Troubleshooting Network Layer 3—Route Tracing tracert, NAT

Task Goal: Be able to identify and troubleshoot Layer 3 points of failure in a network.

The case study of Project 3 is analyzed here, and the task topology is shown in Figure 5.16. Troubleshooting is required since the client's PCs are unable to use the network due to the network devices being misconfigured.

Troubleshooting Network Layer 3— Route Tracing tracert, NAT

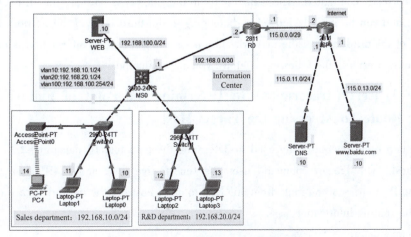

Figure 5.16 Mission Network Topology

1. Check Whether the Client PC Address Information Is Configured Correctly and Whether the Connectivity To the Gateway Is Normal

First, query the IP address information of the client through the ipconfig command, and then use the ping command to detect whether the connectivity from the PC to the gateway is normal or not. The result of checking the IP address information of the client is shown in Figure 5.17, and the result of the connectivity between the client PC and the gateway is shown in Figure 5.18.

```
Packet Tracer PC Command Line 1.0
C:\>ipconfig

FastEthernet0 Connection:(default port)

    Link-local IPv6 Address.........: FE80::201:63FF:FE2A:C2A2
    IP Address......................: 192.168.10.11
    Subnet Mask.....................: 255.255.255.0
    Default Gateway.................: 192.168.10.1
```

Figure 5.17 View Client IP Address Information

```
C:\>ping 192.168.10.1

Pinging 192.168.10.1 with 32 bytes of data:

Reply from 192.168.10.1: bytes=32 time<1ms TTL=255
Reply from 192.168.10.1: bytes=32 time<1ms TTL=255
Reply from 192.168.10.1: bytes=32 time=1ms TTL=255
Reply from 192.168.10.1: bytes=32 time<1ms TTL=255

Ping statistics for 192.168.10.1:
    Packets: Sent = 4, Received = 4, Lost = 0 (0% loss),
Approximate round trip times in milli-seconds:
    Minimum = 0ms, Maximum = 1ms, Average = 0ms
```

Figure 5.18 Check Client PC Connectivity To the Gateway

Analyzing Figure 5.17 and Figure 5.18, it can be seen that the address information of the client PC is correctly configured. The connectivity to the gateway is normal, but still can not access the Internet, and can be initially judged as belonging to the third layer of the network failure. The failure point of the third layer of the network often occurs in the routing information is incomplete or NAT configuration errors, the following is step-by-step elimination.

2. Use Tracert (traceroute) to Determine the Point of Failure of An IP Packet En Route To Accessing the Target Host

Tracert is a route-tracing utility used to determine the path that IP datagrams take to access a destination host. The Tracert command uses the time-to-live field and ICMP error messages to determine the IP address of the route from one host to other hosts on the network as a way to be used for tracing the routing information.

The commands are as follows:

```
C:\>tracert Destination host IP //Route tracing of ICMP packets sent
to destination hosts
```

The results returned when the network is normal are shown in Figure 5.19.

```
C:\>tracert 115.0.13.10

Tracing route to 115.0.13.10 over a maximum of 30 hops:

  1    0 ms    0 ms    0 ms    192.168.10.1
  2    0 ms    0 ms    0 ms    192.168.0.2
  3    2 ms    1 ms    1 ms    115.0.0.2
  4    0 ms    1 ms    0 ms    115.0.13.10

Trace complete.
```

Figure 5.19 Results Returned by Tracert When the Network Is Normal

As you can see from the figure, the packet is sent to the gateway, then to the intranet port of router R0, then to the ISP's gateway, and finally to the target host 115.0.13.10. Each network node will receive an answer and record it.

However, network communication is not normal here, andTracerting the target host on the client PC returns an abnormal result is shown in Figure 5.20.

```
C:\>tracert 115.0.13.10

Tracing route to 115.0.13.10 over a maximum of 30 hops:

  1    0 ms    1 ms    1 ms    192.168.10.1
  2    0 ms    *       0 ms    192.168.10.1
  3    *       0 ms    *       Request timed out.
  4    8 ms    *       0 ms    192.168.10.1
  5    *       1 ms    *       Request timed out.
  6    0 ms    *       0 ms    192.168.10.1
  7    *       0 ms    *       Request timed out.
  8    1 ms    *       0 ms    192.168.10.1
  9    *       1 ms    *       Request timed out.
```

Figure 5.20 Results Returned by Tracert on Network Exceptions

How should the diagnosis be made?

First, analyzing the output information, it can be seen that the data is unreachable after it reaches the gateway, then re-sent to the gateway, and then timeout, and so on and so forth, and it is initially judged that there is an error in the routing table of the switch MS0.

Then, show the routing table on the switch with the following command:

```
MS0#show ip route //view routing table for Layer 3 switch
```

As seen in the routing table for switch MS0 shown in Figure 5.21, switch MS0's routing table has only directly connected routes and no routes to reach the destination network.

Next, check the configuration file of the MS0 of the switch to check whether the routing function of the MS0 of the Layer 3 switch is turned on and whether the routing protocol is correctly configured, and the commands for checking are as follows:

```
MS0#show ip route
Codes: C - connected, S - static, I - IGRP, R - RIP, M - mobile, B - BGP
       D - EIGRP, EX - EIGRP external, O - OSPF, IA - OSPF inter area
       N1 - OSPF NSSA external type 1, N2 - OSPF NSSA external type 2
       E1 - OSPF external type 1, E2 - OSPF external type 2, E - EGP
       i - IS-IS, L1 - IS-IS level-1, L2 - IS-IS level-2, ia - IS-IS inter area
       * - candidate default, U - per-user static route, o - ODR
       P - periodic downloaded static route

Gateway of last resort is not set

     192.168.0.0/30 is subnetted, 1 subnets
C       192.168.0.0 is directly connected, FastEthernet0/24
C    192.168.10.0/24 is directly connected, Vlan10
C    192.168.20.0/24 is directly connected, Vlan20
```

Figure 5.21　Routing Table for Switch MS0

`MS0#show run //view the configuration of the Layer 3 switch`

The configuration file information of switch MS0 is shown in Figure 5.22. ① indicates that the switch has enabled the routing function, which is correct; ② indicates that the switch has configured a default route, and next hop points to the IP address of 192.168.0.254. Comparing with the network topology diagram, it is found that the IP address of next hop is written incorrectly, which is the failure point.

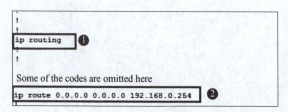

Figure 5.22　Configuration File Information for Switch MS0

Finally, delete the incorrect routing information and add a new routing entry with the following command:

`MS0(config)#ip route 0.0.0.0 0.0.0.0 192.168.0.254 //delete wrong default route`
`MS0(config)#ip route 0.0.0.0 0.0.0.0 192.168.0.2 //Set default route next hop to 192.168.0.2`

After modifying the incorrect default route, revisiting the switch's routing table shows the correct next-hop address, and the correct routing table for switch MS0 is shown in Figure 5.23.

3. Resolving Router NAT Failures

After correcting the routing entry for Layer 3 switch MS0, clients still cannot access the Internet.

First, it is found through Tracert route tracing as in step 2 that the packet has been able to reach the port on router R0 correctly, but it cannot reach the next hop, as shown in Figure 5.24.

```
MS0#show ip route
Codes: C - connected, S - static, I - IGRP, R - RIP, M - mobile, B - BGP
       D - EIGRP, EX - EIGRP external, O - OSPF, IA - OSPF inter area
       N1 - OSPF NSSA external type 1, N2 - OSPF NSSA external type 2
       E1 - OSPF external type 1, E2 - OSPF external type 2, E - EGP
       i - IS-IS, L1 - IS-IS level-1, L2 - IS-IS level-2, ia - IS-IS inter area
       * - candidate default, U - per-user static route, o - ODR
       P - periodic downloaded static route

Gateway of last resort is 192.168.0.2 to network 0.0.0.0

     192.168.0.0/30 is subnetted, 1 subnets
C       192.168.0.0 is directly connected, FastEthernet0/24
C    192.168.10.0/24 is directly connected, Vlan10
C    192.168.20.0/24 is directly connected, Vlan20
S*   0.0.0.0/0 [1/0] via 192.168.0.2
```

Figure 5.23 Correct Routing Table for Switch MS0

```
C:\>tracert 115.0.13.10

Tracing route to 115.0.13.10 over a maximum of 30 hops:

  1    1 ms     0 ms     0 ms  192.168.10.1
  2    1 ms     1 ms     0 ms  192.168.0.2
  3     *        *        *     Request timed out.
  4     *        *        *     Request timed out.
  5     *        *        *     Request timed out.
  6     *        *        *     Request timed out.
  7     *        *        *     Request timed out.
  8     *        *        *     Request timed out.
  9     *        *        *     Request timed out.
```

Figure 5.24 Re-Tracert Route Trace

Check the routing table on router R0, as shown in Figure 5.25, at ① is the static route for the router to return to the intranet VLAN 10, VLAN 20, and at ② is the default route for the router to reach the ISP gateway, and the routing information and the next-hop address are correct.

```
R0#show ip route
Codes: C - connected, S - static, I - IGRP, R - RIP, M - mobile, B - BGP
       D - EIGRP, EX - EIGRP external, O - OSPF, IA - OSPF inter area
       N1 - OSPF NSSA external type 1, N2 - OSPF NSSA external type 2
       E1 - OSPF external type 1, E2 - OSPF external type 2, E - EGP
       i - IS-IS, L1 - IS-IS level-1, L2 - IS-IS level-2, ia - IS-IS inter area
       * - candidate default, U - per-user static route, o - ODR
       P - periodic downloaded static route

Gateway of last resort is 115.0.0.2 to network 0.0.0.0

     115.0.0.0/29 is subnetted, 1 subnets
C       115.0.0.0 is directly connected, FastEthernet1/0
     192.168.0.0/30 is subnetted, 1 subnets
C       192.168.0.0 is directly connected, FastEthernet0/0
S    192.168.10.0/24 [1/0] via 192.168.0.1    ❶
S    192.168.20.0/24 [1/0] via 192.168.0.1
S*   0.0.0.0/0 [1/0] via 115.0.0.2    ❷
```

Figure 5.25 Routing Table for Router R0

Meanwhile, pinging the carrier's gateway address on router R0 is normal, as shown in Figure 5.26.

```
R0#ping 115.0.0.2

Type escape sequence to abort.
Sending 5, 100-byte ICMP Echos to 115.0.0.2, timeout is 2 seconds:
!!!!!
Success rate is 100 percent (5/5), round-trip min/avg/max = 0/0/1 ms
```

Figure 5.26 Detecting Router R0 to ISP Gateway Connectivity

Next, using the knowledge learned in Project 3 to analyze, it is known that the way to achieve access to the Internet server for intranet users is to use Port Address Translation (PAT) technology. After troubleshooting the router's routing configuration and connectivity issues, it is possible to determine that the point of failure is in NAT.

As shown in Figure 5.27, viewing the configuration information on router R0 reveals four error points:

```
R0#show run
Building configuration...
!
! Some of the codes are omitted here
!
interface FastEthernet0/0
 ip address 192.168.0.2 255.255.255.252
 ip nat outside      ①
 duplex auto
 speed auto
!
interface FastEthernet0/1
 no ip address
 duplex auto
 speed auto
 shutdown
!
interface FastEthernet1/0
 ip address 115.0.0.1 255.255.255.248
 ip nat inside       ②
!
interface Vlan1
 no ip address
 shutdown                ③              ④
!
ip nat inside source list 2 interface FastEthernet0/0 overload
ip nat inside source static tcp 192.168.20.13 8080 115.0.0.3 80
ip classless
ip route 192.168.20.0 255.255.255.0 192.168.0.1
ip route 192.168.10.0 255.255.255.0 192.168.0.1
ip route 0.0.0.0 0.0.0.0 115.0.0.2
!
ip flow-export version 9
!
!
access-list 1 permit 192.168.20.0 0.0.0.255
access-list 1 permit 192.168.10.0 0.0.0.255
```

Figure 5.27 Configuration Fault Points for Router R0

At ①, port FastEthernet0/0 is an intranet port and is incorrectly configured as an extranet port.

At ②, port FastEthernet1/0 is supposed to be an external network port and is incorrectly configured as an internal network port.

At ③ is the number of the ACL definition with the latter, which is incorrectly written as 2.

At ④ is that port FastEthernet0/0 is not a port connected to the external network.

These four points of failure are all mistakes that beginners are more likely to make.

Then, against the error point, delete the original wrong configuration and re-correct the configuration of Router R0. The correctly configured code is shown in Figure 5.28.

```
R0#show run
Building configuration...
!
! Some of the codes are omitted here
!
!
interface FastEthernet0/0
 ip address 192.168.0.2 255.255.255.252
 ip nat inside
 duplex auto
 speed auto
!
interface FastEthernet0/1
 no ip address
 duplex auto
 speed auto
 shutdown
!
interface FastEthernet1/0
 ip address 115.0.0.1 255.255.255.248
 ip nat outside
!
interface Vlan1
 no ip address
 shutdown
!
ip nat inside source list 1 interface FastEthernet1/0 overload
ip nat inside source static tcp 192.168.20.13 8080 115.0.0.3 80
ip classless
ip route 192.168.20.0 255.255.255.0 192.168.0.1
```

Figure 5.28 R0 Configuration of Router after Fixing Failure

Finally, accessing the Internet via Tracert trace on the client PC and using a browser to access the baidu server have been successful, as shown in Figure 5.29 and Figure 5.30.

```
C:\>tracert 115.0.13.10

Tracing route to 115.0.13.10 over a maximum of 30 hops:

  1    0 ms    1 ms    0 ms    192.168.10.1
  2    0 ms    1 ms    0 ms    192.168.0.2
  3    2 ms    0 ms    0 ms    115.0.0.2
  4    0 ms    0 ms    0 ms    115.0.13.10

Trace complete.
```

Figure 5.29 Client PC access to the Internet using Tracert tracing

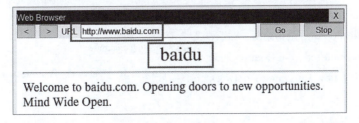

Figure 5.30 Client PC Accessing Baidu Services Using A Browser

Exercise: complete the three-layer fault network model that accompanies the eResource Library to realize that intranet users can access Internet resources.

Competence Expansion: Layer 3 network faults through Tracert, ping and other commands, combined with a variety of tools for fault location and troubleshooting, you can quickly accumulate troubleshooting experience and improve the technical level. We should be good at using various tools to improve our efficiency in solving various problems. Marxism reveals the essence of things, the inner connection and the law of development, is the "great tool of understanding", which is a powerful ideological weapon for people to observe the world and analyze problems. Only by establishing the scientific faith of Marxism can college students truly establish lofty ideals and beliefs, group the essence of the intricate social phenomena, clarify the direction, and make greater contributions to serving the people and contributing to the society.

Summary and Expansion

1. Network Troubleshooting Lessons Learned

Network failures generally fall into two broad categories: connectivity problems and performance problems.

(1) Connectivity issues.

It mainly appears in hardwares, systems, power supply, transmission media failures, IP address misconfigurations and so on.

(2) Performance issues.

It mainly occurs in network congestion, to destinations that are not best paths, forwarding anomalies, routing loops, and so on.

Troubleshooting involves a step-by-step process of identifying the cause of a fault and solving it, and it is based on the idea of systematical dividing (or segregating) a large set of possible causes of a fault into several smaller subsets, thus rapidly reducing the complexity of the problem. An organized mindset of troubleshooting helps to resolve any difficulties encountered, and Figure 5.31 gives the processing flow for general network troubleshooting.

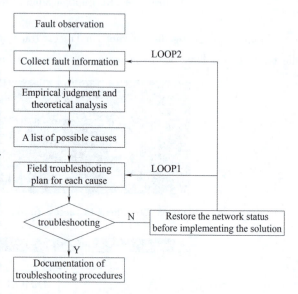

Figure 5.31 Network Troubleshooting Process

2. Summary of Network Troubleshooting Commands for Everyday Use

(1) ping.

In network troubleshooting, the most frequently-used is the ping command, which not only checks whether the network is connected, but also helps to analyze and determine network faults. Its common methods are:

①ping local IP.

The machine should always respond to this ping command, and if it does not, there is a problem with the local TCP/IP stack.

②ping gateway IP.

You can check if the machine is connected to the local network properly.

③Ping the remote IP.

If 4 responses are received, the network is normal.

④Ping other IPs on the LAN.

The packets will pass through the network card and the network cable to other machines before returning. Receiving a return answer indicates correct operation in the local network; however, if you receive zero return answers, then the subnet mask is incorrect or the network card is misconfigured or there is a problem with the network cable.

⑤Ping 127.0.0.1.

The loopback address is to check that the local TCP/IP protocol is set up correctly.

⑥pinq www.xxx.com (e.g. www.baidu.com).

Check whether DNS can resolve the correct IP address and reach the target server properly.

(2) tracert.

tracert is designed to probe the path that data packets pass through between the source node and the destination node. If there is a network connectivity problem, you can use the tracert command to check the path to the destination IP address that is reached and records the result. The tracert command displays the set of IP routers used to deliver the packet from the computer to the destination location and the time taken at each leap point. If the packet cannot be delivered to the destination, the tracert command displays the last router that successfully forwards the packet. The tracert command can be used to trace the routes (paths) used in the transmission of data when datagrams are transmitted from our computers to their destinations through multiple gateways. The tracert command is very simple to use Following the tracert with an IP Tracert is very simple to use, you just need to follow the tracert with an IP address or a domain name to determine which part of the network has problems.

Thinking and Training

1. Short answer questions

(1) What are the common failures of LAN? What are the possible reasons for the faults?

(2) What are some common network commands for network troubleshooting? What is the

function of each command?

(3) What are the main methods of LAN troubleshooting?

(4) How do the steps of analyzing a network when it fails? Please describe in your own words.

2. Experiment questions

Complete the troubleshooting network model for the companion electronic repository project 5, which will ultimately enable intranet users to access Internet resources while allowing external network users to access the intranet servers.

参 考 文 献

[1] [美] 里克·格拉西亚尼，艾伦·约翰逊. 思科网络技术学院教程：网络基础 [M]. 田果，译. 北京：人民邮电出版社，2023.

[2] 思科网络技术学院. 思科网络技术学院教程网络简介实验手册 [M]. 北京：人民邮电出版社，2015.

[3] 邓启润，钟文基. 网络互联技术任务驱动式教程 [M]. 北京：人民邮电出版社，2021.

[4] 严争. 计算机网络基础教程 [M]. 北京：电子工业出版社，2016.

[5] [美] 里克·格拉齐亚尼. 思科网络技术学院教程 [M]. 思科系统公司，译. 北京：人民邮电出版社，2022.